ANALYTICAL CHEMISTRY FROM LABORATORY TO PROCESS LINE

ANALYTICAL CHEMISTRY FROM LABORATORY TO PROCESS LINE

Edited by
Gennady E. Zaikov, DSc, and A. K. Haghi, PhD

Apple Academic Press Inc. | Apple Academic Press Inc.
3333 Mistwell Crescent | 9 Spinnaker Way
Oakville, ON L6L 0A2 | Waretown, NJ 08758
Canada | USA

© 2016 by Apple Academic Press, Inc.

First issued in paperback 2021

Exclusive worldwide distribution by CRC Press, a member of Taylor & Francis Group

No claim to original U.S. Government works

ISBN 13: 978-1-77463-424-0 (pbk)
ISBN 13: 978-1-77188-735-9 (hbk)

Library and Archives Canada Cataloguing in Publication

Analytical chemistry from laboratory to process line / edited by Gennady E. Zaikov, DSc, and A. K. Haghi, PhD.

Includes bibliographical references and index.
Issued in print and electronic formats.
ISBN 978-1-77188-735-9 (hardcover).--ISBN 978-0-429-45435-6 (pdf)
1. Chemical engineering--Research. 2. Chemistry, Analytic. I. Haghi, A. K author, editor
II. Zaikov, G. E. (Gennadiï Efremovich), 1935-, author, editor

| TP155.A53 2015 | 660 | C2015-906455-4 | C2015-906456-2 |

Library of Congress Cataloging-in-Publication Data

Analytical chemistry from laboratory to process line / [edited by] Gennady E. Zaikov, DSc, and A.K. Haghi, PhD.

pages cm
Includes bibliographical references and index.
ISBN 978-1-77188-735-9 (alk. paper)--ISBN 978-0-429-45435-6 (ebook)
1. Chemical engineering. 2. Chemistry, Analytic. 3. Materials--Research. I. Zaikov, G. E. (Gennadii Efremovich), 1935- II. Haghi, A. K.

| TP155 .A53 | 660--dc23 | 2015035981 |

Apple Academic Press also publishes its books in a variety of electronic formats. Some content that appears in print may not be available in electronic format. For information about Apple Academic Press products, visit our website at **www.appleacademicpress.com** and the CRC Press website at **www.crcpress.com**

CONTENTS

LIST OF CONTRIBUTORS

Arezo Afzali
University of Guilan, Rasht, Iran

T. I. Aksenova
Associated Professor, Moscow State University of Technologies and Management named after K.G. Razumovskiy, 73, Zemlyanoy Val St., Moscow, Russia; E-mail: aksentatyana@rambler.ru

Kuvshinova Larisa Alexandrovna
Federal State Institution of Science, Institute Chemistry of Komi Scientific Centre of the Ural Branch of the Russian Academy of Sciences, Pervomaiskaya St., 48, Syktyvkar, 167982, Russia, E-mail: fragl74@mail.ru

B. S. Alikhadzhieva
Chechen State Pedagogical Institute, 33, Kievskaj Str., 364037, Grozny, Russia; E-mail: belkaas52@list.ru

Dzhankulaeva Madina Amerhanovna
Kabardino-Balkarian State University of H.M. Berbekov, Moscow, Russia

V. V. Ananyev
Head of Laboratory, Research Center of Ivan Fedorov Moscow State University of Printing Arts, 2A, Pryanishnikova St., Moscow, Russia; E-mail: vovan261147@rambler.ru

N. T. Arlamova
Dneprodzerzhynsk State Technology University, Ukraine

Yu. V. Berestneva
Donetsk National University, Universitetskaya Street, 24, Donetsk, 83 001, Ukraine

T. A. Borukaev
Kabardino-Balkarian State University 173, Chernyshevsky Str., 360004, Nalchik, Russia

N. A. Bublyk
Institute of Horticulture, NAAS of Ukraine, Kiev, Ukraine

A. I. Burya
Dneprodzerzhynsk State Technology University, Ukraine

A. T. Djalilov
State Unitary Enterprise Tashkent Research, Institute of Chemical Technology, Tashkent, Uzbekistan

I. S. Eremeev
A.V. Topchiev Institute of Petrochemical Synthesis, Russian Academy of Sciences, 29, Leninsky prospect, 119991, Moscow, Russia

Vladimir S. Feofanov
N.M. Emanuel Institute of Biochemical Physics of the Russian Academy of Sciences, Kosygin St. 4 117977 Moscow, Russia

A. K. Haghi
University of Guilan, Raht, Iran

A. A. Kambarova
State Unitary Enterprise Tashkent Research, Institute of Chemical Technology, Tashkent, Uzbekistan

M. U. Karimov
State Unitary Enterprise Tashkent Research, Institute of Chemical Technology, Tashkent, Uzbekistan

G. P. Karpacheva
A.V. Topchiev Institute of Petrochemical Synthesis, Russian Academy of Sciences, 29, Leninsky prospect, 119991, Moscow, Russia

Z. S. Khasbulatova
Chechen State Pedagogical Institute, 33, Kievskaj Str., 364037, Grozny, Russia; E-mail: belkaas52@list.ru

S. G. Kiseleva
TIPS RAS, Moscow, Russia

O. I. Kitaev
Institute of Horticulture, NAAS of Ukraine, Kiev, Ukraine

N. A. Kleshcheva
A.V. Topchiev Institute of Petrochemical Synthesis, RAS (TIPS RAS), Moscow, Russia

Lubov Kh. Komissarova
N.M. Emanuel Institute of Biochemical Physics of the Russian Academy of Sciences, Kosygin St. 4 117977 Moscow, Russia, Tel.: 8(495)9361745 (office), 8(906)7544974 (mobile); Fax: (495)1374101; E-mail: komissarova-lkh@mail.ru

G. A. Korablev
Izhevsk State Agricultural Academy, Studencheskaya St., 11, Izhevsk, 426000, Russia, E-mail: korablevga@mail.ru

G. V. Kozlov
Kh.M. Berbekov Kabardino-Balkarian State University, Chernyshevsky St., 173, 360004 Nalchik, Russian Federation; E-mail: i_dolbin@mail.ru

A. M. Kugotova
Kh.M. Berbekov Kabardino-Balkarian State University, Nalchik, Russia; E-mail: kam-02@mail.ru

P. P. Kulikov
Graduate Student, D. Mendeleyev University of Chemical Technology of Russia, 9, Miusskaya sq., Moscow, Russia; E-mail: p.kulikov.p@gmail.com

B. I. Kunizhev
Kh.M. Berbekov Kabardino-Balkarian State University, Nalchik, Russia; E-mail: kam-02@mail.ru

G. Levin
Research Institute for Antioxidant Therapy, 137c Invalidenstr., 10115 Berlin, Germany

M. M. Ligidova
Kabardino-Balkarian State University 173, Chernyshevsky Str., 360004, Nalchik, Russia

Shima Maghsoodlou
University of Guilan, Rasht, Iran

A. I. Martynenko
A.V. Topchiev Institute of Petrochemical Synthesis, RAS (TIPS RAS), Moscow, Russia

M. R. Menyashev
A.V. Topchiev Institute of Petrochemical Synthesis, RAS (TIPS RAS), Moscow, Russia

A. K. Mikitaev
Kabardino-Balkarian State University, 173, Chernyshevsky Str., 360004, Nalchik, Russia

M. M. Murzakanova
Kabardino-Balkarian State University, 173, Chernyshevsky Str., 360004, Nalchik, Russia;
E-mail: m_m_murzakanova@mail.ru

Oshkhunov Muaed Muzafarovich
Kabardino-Balkarian State University of H.M. Berbekov, Moscow, Russia

O. A. Naberezhnaya
Dneprodzerzhynsk State Technology University, Ukraine

Manahova Tat'yana Nicolaevna
Northern (Arctic) Federal University named after M.V. Lomonosov, Northern Dvina Emb., 17,
Arkhangelsk, 163002, Russia, E-mail: tatiankaya17@yandex.ru

Babak Noroozi
University of Guilan, Rasht, Iran

A. V. Orlov
TIPS RAS, Moscow, Russia

S. Zh. Ozkan
A.V. Topchiev Institute of Petrochemical Synthesis, Russian Academy of Sciences, 29, Leninsky
prospect, 119991, Moscow, Russia

N. V. Patyka
NSC "Institute of Agriculture of NAAS of Ukraine," Kiev, Ukraine

T. I. Patyka
Institute of Horticulture, NAAS of Ukraine, Kiev, Ukraine

I. N. Popov
Research Institute for Antioxidant Therapy, 137c Invalidenstr., 10115 Berlin, Germany,
E-mail: ip@antioxidant-research.com

N. I. Popova
A.V. Topchiev Institute of Petrochemical Synthesis, RAS (TIPS RAS), Moscow, Russia

E. V. Raksha
Donetsk National University, Universitetskaya Street, 24, Donetsk, 83 001, Ukraine

N. A. Samigov
Tashkent Architectural and Construction Institute, Tashkent, Uzbekistan

N. N. Sazhina
Emanuel Institute of Biochemical Physics Russian Academy of Sciences, 4 Kosygin Street, 119334
Moscow, Russia, E-mail: Natnik48 s@yandex.ru

N. A. Sivov
A.V. Topchiev Institute of Petrochemical Synthesis, RAS (TIPS RAS), Moscow, Russia

V. D. Tretyakova

Project Manager, METACLAY CJSC, 15, Karl Marx St., Karachev, Bryansk Region, Russia; E-mail: vera.d.tretyakova@gmail.com

G. G. Tukhtaeva

State Unitary Enterprise Tashkent Research, Institute of Chemical Technology, Tashkent, Uzbekistan

N. A. Turovskij

Donetsk National University, Universitetskaya Street, 24, Donetsk, 83 001, Ukraine; E-mail: na.turovskij@gmail.com; elenaraksha411@gmail.com

R. R. Usmanova

Ufa State Technical University of Aviation, 12 Karl Marks Str., Ufa 450100, Bashkortostan, Russia; E-mail: Usmanovarr@mail.ru

G. S. Valiullina

Izhevsk State Agricultural Academy, Studencheskaya St., 11, Izhevsk, 426000, Russia

G. E. Zaikov

N.M. Emanuel Institute of Biochemical Physics, Russian Academy of Sciences, 4 Kosygin Str., Moscow 119334, Russia; E-mail: chembio@sky.chph.ras.ru

LIST OF ABBREVIATIONS

AA	acrylic acid
AFM	atomic force microscopy
AOA	total antioxidant activity
ASC	ascorbic acid
BNCT	boron neutron capture of tumor therapy
CNT	carbon nanotube
CP	cyclic products
DCV-GCMD	dual-volume GCMD
DEL	double electric layer
DP	diamond pore
DPA	diphenylamine
FEM	finite element method
GCMD	grand canonical molecular dynamics
GGH	guanidine hydrochloride
HDPE	high density polyethylene
IUPAC	international union of pure and applied chemistry
L-BPA	L-borophenilalanin
LDPE	low-density polyethylene grade
MAA	methacrylic acid
MAAIBE	methacrylic acid isobutyl ester
MAG	methacrylate guanidine
MC	Monte Carlo
MD	molecular dynamics
MF	microfiltration
MGHC	methacryloyl guanidine hydrochloride
MGU	methacryloyl guanidine
MMT	montmorillonite
MSD	mean-square displacement
MWCO	molecular weight cut-off
MWNT	multi-walled carbon nanotube

NCT	neutron capture therapy
NEMD	non-equilibrium MD
NF	nanofiltration
OGB	oxide tungsten bronzes
OP	organic plastics
PCL	photo-sensitized
PCM	continuum polarization model
PMMA	poly(methyl methacrylate)
PP	polypropylene
PPhOA	polyphenoxazine
PS	polystyrene
PVC	poly(vinyl chloride)
RESPA	reference system propagator algorithm
RO	reverse osmosis
ROOH	1,1,3-trimethyl-3-(4-methylphenyl)butyl hydroperoxide
SP	straight path
SSS	stress-strain state
SWNT	single-walled carbon nanotube
TGA	thermal gravimetric analysis
TIC	thermo-initiated CL
TMS	tetramethylsilane
TRAP	total antioxidant potencial
UA	uric acid
UF	ultrafiltration
ZP	zigzag path

LIST OF SYMBOLS

A	proportionality constant
a_1 and a_2	unit vectors
aq	energy barrier
B	hole affinity constant
$c(x)$	concentration
c_A	concentration of diffusant A
C_h^*	saturation constant
D	diffusivity (diffusion coefficient)
D	pore diameter
D_A^*	self-diffusion coefficient
e_1, e_2 and e_3	coordinates in the current configuration
F	applied external force
f_A	fugacity
G_1, G_2	material coordinates of a point in the initial configuration
J	molecular flux
K	temperature dependent Henry's law coefficient
K_n	Knudsen number
K_0	proportionality constant
K_P	Henry's law constant
L	membrane thickness
M	molecular mass
N	surface concentration of pores
Na	number of molecules
P	permeability
P_S^*	constant
Q	flow
Q	heat of adsorption
R	radius of the modeled SWCNT
r_p	pore radius
S	solubility coefficient

T	number of trials
T	pore tortuosity
T	thickness of the adsorbate film
T	time
V	center-of-mass velocity component
V_L	molecular volume of the condensate
X	position across the membrane

Greek Symbols

ξ_i	random numbers generated for each trial
Γ	surface tension
Δp	pressure drop across the membrane
Θ	time-lag
Λ	mean free path of molecules
$\rho(x)$	an arbitrary probability distribution function
$\bar{u}$	average molecular speed

PREFACE

The main attention in this book of collected scientific papers is paid to the recent theoretical and practical advances in engineering chemistry research.

This volume highlights the latest developments and trends in engineering chemistry research. It presents the developments of advanced Engineering Chemistry Research and respective tools to characterize and predict the material properties and behavior. The book is aimed at providing original, theoretical, and important experimental results that use non-routine methodologies often unfamiliar to the usual readers. Furthermore chapters on novel applications of more familiar experimental techniques and analyses of composite problems indicate the need for new experimental approaches presented.

Technical and technological development demands the creation of new materials, which are stronger, more reliable and more durable, i.e. materials with new properties. Up-to-date projects in the creation of new materials go along the way of engineering chemistry research.

The technology of engineering chemistry research forges ahead; its development is directed to the simplification and cheapening the production processes of composite materials with new particles in their structure. However, the engineering chemistry research develops at high rates; what seemed impossible yesterday, will be accessible to the introduction on a commercial scale tomorrow.

The desired event of fast implementation of engineering chemistry research in mass production depends on the efficiency of cooperation between scientists and manufacturers in many respects. Today's high technology problems of applied character are successfully solved in close consolidation of the scientific and business worlds.

With contributions from experts from both industry and academia, this book presents the latest developments in the identified areas. This book incorporates appropriate case studies, explanatory notes, and schematics for clarity and understanding.

This book will be useful for chemists, chemical engineers, technologists, researchers, and students interested in advanced engineering chemistry research with complex behavior and their applications

On the other hand, this book is also designed to fulfill the requirements of scientists and engineers who wish to be able to carry out theoretical and experimental research in engineering chemistry research using modern methods. Each chapter in section one describes the principle of the respective method as well as the detailed procedures of experiments, with examples of actual applications presented in section two. Thus, readers will be able to apply the concepts as described in the book to their own experiments.

Experts in each of the areas covered have reviewed the state of the art, thus creating a book that will be useful to readers at all levels in academic, industry, and research institutions.

Engineers, polymer scientists, and technicians will find this volume useful in selecting approaches and techniques applicable to characterizing molecular, compositional, rheological, and thermodynamic properties of elastomers and plastics.

ABOUT THE EDITORS

Gennady E. Zaikov, DSc
Gennady E. Zaikov, DSc, is Head of the Polymer Division at the N. M. Emanuel Institute of Biochemical Physics, Russian Academy of Sciences, Moscow, Russia, and Professor at Moscow State Academy of Fine Chemical Technology, Russia, as well as Professor at Kazan National Research Technological University, Kazan, Russia. He is also a prolific author, researcher, and lecturer. He has received several awards for his work, including the Russian Federation Scholarship for Outstanding Scientists. He has been a member of many professional organizations and on the editorial boards of many international science journals.

A. K. Haghi, PhD
A. K. Haghi, PhD, holds a BSc in urban and environmental engineering from University of North Carolina (USA); a MSc in mechanical engineering from North Carolina A&T State University (USA); a DEA in applied mechanics, acoustics and materials from Université de Technologie de Compiègne (France); and a PhD in engineering sciences from Université de Franche-Comté (France). He is the author and editor of 165 books as well as 1000 published papers in various journals and conference proceedings. Dr. Haghi has received several grants, consulted for a number of major corporations, and is a frequent speaker to national and international audiences. Since 1983, he served as a professor at several universities. He is currently Editor-in-Chief of the *International Journal of Chemoinformatics and Chemical Engineering* and *Polymers Research Journal* and on the editorial boards of many international journals. He is a member of the Canadian Research and Development Center of Sciences and Cultures (CRDCSC), Montreal, Quebec, Canada.

PART 1

APPLIED CHEMISTRY
RESEARCH NOTES

CHAPTER 1

A NOTE ON VAN DER WAALS RADII AND ATOMIC DIMENSIONAL CHARACTERISTICS

G. A. KORABLEV,[1] G. S. VALIULLINA,[1] and G. E. ZAIKOV[2]

[1]Izhevsk State Agricultural Academy, Studencheskaya St., 11, Izhevsk, 426000, Russia, E-mail: korablevga@mail.ru

[2]Institute of Biochemical Physics N.M. Emanuel, 4, Kosygin St., Moscow, 119991, Russia, E-mail: chembio@sky.chph.ras.ru

CONTENTS

ABSTRACT

The registration of P-parameter properties allows explaining the direct dependence between covalence and Van der Waals radii.

1.1   INTRODUCTION

Covalence and Van der Waals radii are widely used in physical-chemical research.

The covalence bond is the bond formed by a pair of electrons. At the same time, each atom included into the bond sends one electron to the pair, which belongs to both atoms.

Van der Waals interactions are intermolecular interactions between electrically neutral particles. These are weak interactions but they are very important in structural conformation processes, especially for biosystems.

However, there is still no mathematical bond between the radii of these significant interactions. The concept of spatial-energy parameter (P-parameter) is applied for this purpose in this research [1].

1.2   RESEARCH METHODS

It was found that P-parameter possesses wave properties [1].

The interference maximum, amplification of oscillations (in the phase) occur if the wave path difference equals the even number of half-waves:

$\Delta = 2n\dfrac{\lambda}{2} = \lambda n$ or $\Delta = \lambda(n+1)$, where λ – wavelength, n = 0, 1, 2, 3...
(even number).

As applicable to P-parameter, the maximum amplification of interactions in the phase corresponds to the interactions of like-charged systems or systems homogeneous by their properties and functions (e.g., between the fragments or blocks of complex organic structures). Then, the relative value of P-parameters of these systems is as follows:

$$\gamma = \frac{P}{P_i} = (n+1) \tag{1}$$

Similarly, for "degenerated" systems (with similar function values) of 2D harmonic oscillator, the energy of stationary states is as follows:

$$\varepsilon = hv(n+1),$$

where h – Plank's constant, v – frequency.

From the Eq. (1) we have: $P_i = \dfrac{P}{n+1}$ or $P_i = \dfrac{P}{n}$.

This means that apart from the initial (main) atom state with P-parameter, each atom can have structural-active valence orbitals with another value of P_i-parameters, moreover, the nearest most active valence states differ in two times by the values of P-parameters. Formally, this corresponds to the increase in the distance of interatom (intermolecular) interaction in 2 times, that is, we observe the transition from the interaction radius to the diameter.

Therefore, for P_e-parameter we have:

$$P'_e = \frac{P_0}{R_1(n+1)} \qquad (2)$$

or

$$P''_e = \frac{P_0}{R_2 n} \qquad (3)$$

Apparently, the periodicity of element system also corresponds to the Eqs. (2) and (3), in which, taking into account the screening effects, it is better to use the effective main quantum number – n^* instead of n, the bond between which by Slatter [2] is as follows:

$$n\ 1\ 2\ 3\ 4\ 5\ 6$$

$$n^*1\ 2\ 3\ 3,7\ 4\ 4,2$$

Then: $P'_e = \dfrac{P_i}{R_1(n^*+1)}$ and $P''_e = \dfrac{P_i}{R_2 n^*}$

where $P_i - P_0$-parameters of each element in the given period of the System.

1.3 CALCULATIONS AND COMPARISONS

It is known that an electron energy, if there are no other electrons on the orbital, depends only on $\left(\dfrac{Z^*}{n^*}\right)^2$, where Z^* – nucleus effective charge.

In accordance with the equality principle of P-parameters of interacting systems and as applicable to the given atom at its different radii of inter-molecular interaction, we make the Eqs. (2) and (3) equal, and, raising $n*$ and $(n*+1)$ to the second power, we have:

$$r_k \left(n*+1\right)^2 = R_v n*^2 \rightarrow R_v = \left(\frac{n*+1}{n*}\right)^2 r_k , \qquad (4)$$

where r_k – covalence radii, R_v – Van der Waals radii.

The correctness of the Eq. (4) is proved by the calculations given in Table 1.1. By physical sense, this equation is determined by quantum

TABLE 1.1   Dependence Between Covalence and Van der Waals Radii

Period	$\gamma = \left(\dfrac{n*+1}{n*}\right)$	Atom	r_k(Å)	γr_k (Å)	R_v(Å)
I	$\left(\dfrac{1+1}{1}\right)^2 = 4$	H	0.28	1.120	1.10
II	$\left(\dfrac{2+1}{2}\right)^2 = 2.25$	B	0.80	1.800	1.75
		C	0.77	1.733	1.70
		N	0.70	1.575	1.50
		O	0.66	1.485	1.40
		F	0.64	1.440	1.35
III	$\left(\dfrac{3+1}{3}\right)^2 = 1.778$	Si	1.11	1.974	1.95
		P	1.10	1.956	1.90
		S	1.04	1.849	1.85
		Cl	1.0	1.778	1.80
IV	$\left(\dfrac{3.7+1}{3.7}\right)^2 = 1.6136$	Ga	1.25	2.017	2.0
		Ge	1.24	2.001	2.0
		As	1.21	1.952	2.0
		Se	1.17	1.888	2.0
		Br	1.20	1.936	1.95
V	$\left(\dfrac{4+1}{4}\right)^2 = 1.5625$	Sn	1.40	2.188	2.20
		Sb	1.41	2.203	2.20
		Te	1.37	2.141	2.20
		I	1.35	2.109	2.15

changes in the radius of intermolecular interaction of elements of different periods of the System.

Thus, covalence and Van der Waals radii are linked by a simple dependence via the coefficient $\left(\dfrac{n*+1}{n*} \right)^2$.

KEYWORDS

- covalence
- P-parameter properties
- spatial-energy parameter
- Van der Waals radii

REFERENCES

1. Korablev, G. A. Spatial-Energy Principles of Complex Structures Formation, Amsterdam, The Netherlands, Brill Academic Publishers and VSP, 2005, 426p.
2. Batsanov, S. S., Zvyagina, R. A. Overlap integrals and problem of effective charges (in Rus.), Science (Nauka) Publishing House, Novosibirsk City, 1966, 386p.

CHAPTER 2

A NOTE ON INORGANIC POLYMERS SEMICONDUCTOR MATERIALS

B. S. ALIKHADZHIEVA and Z. S. KHASBULATOVA

Chechen State Pedagogical Institute, 33, Kievskaj Str., 364037, Grozny, Russia; E-mail: belkaas52@list.ru

CONTENTS

ABSTRACT

Metaphosphates alkali metals of inorganic polymers, which in the molten state retain polymer structure as electrolytes with high ionic conductivity and have a substantially viscosity.

Due to introduction to polytungstate systems the phosphates alkali metals the alkaline tungsten bronzes were obtained, which are used as catalysts.

Experimentally state diagrams were investigated involving metaphosphates of sodium (potassium) with tungsten alkali metals (Na, K).

The systems metaphosphate-tungstate sodium and potassium can be used to create the crystallization resistant vitreous semiconductor materials.

2.1   INTRODUCTION

One of the most important problems of modern inorganic chemistry is the obtaining of new polymer and composite materials with predetermined properties. In nowadays, there are high requirements to the quality of oxide and oxide-salt materials such as powders, ceramics, pellicles and fibers have led to the development of principally new methods of obtaining them [1].

Complex oxide tungsten containing phases, with unique physical and chemical properties, are promising inorganic materials for creation of new engineering and technologies. Therefore, considerable attention is paid to the improvement and development of theoretical and practical bases of obtaining them [2].

Currently, one of the main methods of obtaining them is electrolysis of oxide melts and tungsten salts with implementation as thinners the electrolytes more low-melting component. However, there is the formation of macrocrystalline precipitation, but for use, for example, oxide tungsten bronzes (OGB) as catalysts are required powders of high dispersity [3].

One of the possible solutions is to set a high-viscosity melts such as electrolytes, which in particular achieved by the introduction in polytungstate system phosphates alkaline metals [4].

Physicochemical properties vary depending on the amount of metal embedded: color, texture, conductivity, etc. Crystal lattice of the oxide tungsten bronzes is built from octahedra three-tungsten oxide, interconnected in a variety of ways. There are some voids between octahedra where an ion can fit with no distortion of the lattice; the size is equal to or less oxygen. Depending on how connected octahedra of tungsten trioxide with each other, and which kind of voids with the form, we can obtain the structures of one or another crystal or a crystallographic symmetry. In particular, for oxide tungsten bronzes currently known cubic, tetragonal, hexagonal, orthorhombic, monoclinic structure [5].

A wide range of compositions oxide tungsten bronzes opens the possibility to vary with valuable physical and chemical properties. The most studied of all the oxide tungsten bronzes are alkaline tungsten bronze. Research of acid-base properties of melts in the system Na_2WO_4 - $NaPO_3$ also showed that the potential platinum oxygen electrode, immersed in the study melts, is moving to the positive area while increasing them in concentration metaphosphate sodium [6] which, obviously, is also associated with the anionic polymerization groups of tungsten, consequently, reduced activity of oxygen ions:

$$2WO_4^{2-} + PO_3^- \leftrightarrow PO_4^- + W_2O_7^2$$

$$PO_3^- + O_2 \leftrightarrow PO_4^{3-}$$

Anion PO_3^- as a strong acceptor ion of oxygen, with the injection of tungstate melts shifts the reaction to the right, and induces the polymerization tungstate-ions as well as in these melts with the injection the oxide tungsten (VI) [7].

Thereby, injection to tungsten melts of metaphosphate of alkali metal leads to an increase in the melt concentration double tungstate ion $W_2O_7^{2-}$ that "supply" in the melt WO_3 particles, that is, the melt is source of tungsten (VI). Therefore, in melts tungsten phosphate systems in the chemical way it is possible to synthesize in oxide tungsten bronzes powders (OGB), but in the absence of tungsten (VI). Phosphorus has the ability to form the various polymer compounds with the range of properties.

The most complete information about the interaction of condensed phosphates is available by examining the phase diagram of the complex of methods of physical-chemical analysis, including the DTA, the ARF. These data give an idea about the state, the properties of the solid and liquid phases, the areas of glass formation. Such approach allows to avoid unnecessary losses of substances and time when selecting a practical important compositions.

It seemed interesting to study of phase diagrams with the participation of inorganic polymers-metaphosphate sodium (potassium) with tungstate of alkali metals (Na, K), which still not enough studied [2].

Metaphosphates of alkali metals belong to the class of inorganic polymers, which in the molten state retain the polymer structure as electrolytes with high ionic conductivity. One of the most important advantages of metaphosphates of alkali metals with polymeric structure is the ability to dissolve the oxides of many metals. Like many other inorganic polymers, metaphosphates of alkali metals in molten form have considerable viscosity, which is conditioned by the peculiarities of polymer structure of these compounds [2].

According to the data of Ref. [3] the degree of polymerization of metaphosphates of alkaline metal minerals increased in the range: $LiPO_3$, KPO_3. Viscosity according to the data of Ref. [4] with increasing radius cation decreases. At interaction with some oxides, for example, V_2O_5, the metaphosphates form complex ions. The increase in density and viscosity in melts is explained by formation of complex ions. With oxides of some metals, the metaphosphates of alkaline metals are forming the glass.

Metaphosphate of alkaline metals, according to data of Refs. [5, 6] thermally resistant to temperatures of 100–150°C above their melting point.

Initial condensed metaphosphates were received by us with the method of solid-phase reactions of interaction of salt (soda, a potassium) with orthophosphoric acid:

$$H_3PO_4 + Na_2CO_3 = 2NaH_2PO_4 + CO_2 + H_2O$$

$$NaH_2PO_4 = NaPO_3 + H_2O$$

$$H_3PO_4 + K_2CO_3 = 2KH_2PO_4 + CO_2 + H_2O$$

$$KH_2PO_4 = KPO_3 + H_2O$$

2.2   IDENTIFICATION OF PHASES WAS CARRIED OUT BY THE X-RAY PHASE ANALYSIS

The results of the research of interaction of metaphosphate of sodium with sodium tungstate and complex of methods of the physical and chemical analysis allowed to establish that in this $NaPO_3 - Na_2WO_4$ system form two dystectics (D1) with melting point 680°C, (D2) with melting point 570°C, forms four eutectics e_1, e_2, e_3, e_4 with melting points 612°C, 580°C, 540°C, 500°C, respectively.

The $KPO_3 - K_2WO_4$ system is also studied. Its components form congruent connections of $K_2WO_4 - 2KPO_3$ (D$_3$) with melting point 646°C, and to the eutectic points there correspond structures 55 (620°C) and 75 mol. % (618°C) KPO_3.

The chart of fusibility of data of systems was constructed of experimentally obtained data and fields of crystallizing phases and character of nonvariant points are outlined. The greatest ability to formation of double salts metaphosphates of monovalent metals is differ. It should be noted the tendency of metaphosphates mono and divalent metals to formation of restricted solid solutions. Studying of crystallization ability in melts metaphosphate – sodium tungstate and a potassium of data of systems is of interest to creation of crystallization and steady vitreous semiconductor materials.

KEYWORDS

- **electrochemical synthesis**
- **eutectic**
- **oxide tungsten bronzes**
- **phase transitions**
- **two-component systems**

REFERENCES

1. Alikhadzhieva, B. S. Thesis PhD. Makhachkala. 2011, pp. 118.
2. Materials IV of the All Russian Scientific Bergmanovsky Conference with the International Participation. Makhachkala, 13–14 April, 2012.
3. Mandelkorn, L. Nonstoichiometric Compounds. "Khimiya" ("Chemistry", in Rus.) Publishing House, 1971, p. 607.
4. Scheibler, C. Uber wolframoxyd verbindungen. J. Ract. Chem., 1861, B.183, 320–324.
5. Ozerov, R. P. Tungsten and Vanadium Bronzes. "The Reports of Russian Academy of Sciences" (in Rus.), 1954, v. 99, №1. 93–95.
6. Spitsin, V. I. Oxide Bronzes. "Nauka" ("Science", in Rus.) Publishing House, 1982.
7. Kollong, R. Nonstoichiometry. "Mir" Publishing House (in Rus.), 1974, p. 287.

CHAPTER 3

A NOTE ON THE INFLUENCE OF COPOLYMER OF NA-ACRYLIC ACID AND METHACRYLIC ACID ISOBUTYL ESTER ON THE PHYSIC-MECHANICAL PROPERTIES OF CEMENT SYSTEMS

M. U. KARIMOV,[1] A. T. DJALILOV,[1] and N. A. SAMIGOV[2]

[1]State Unitary Enterprise Tashkent Research, Institute of Chemical Technology, Tashkent, Uzbekistan

[2]Tashkent Architectural and Construction Institute, Tashkent, Uzbekistan

CONTENTS

ABSTRACT

This chapter explains synthesis and influence of copolymers of Na-acrylic acid and methacrylic acid isobutyl ester on the physic-mechanical properties of cement systems. Cement systems were investigated, according to state standard specifications. Studied the fluidity of water–cement mortar and strength of a cement stone with the addition of chemical additives.

3.1   INTRODUCTION

In contrast to many modifiers superplasticizer based on polycarboxylate esters are attached to the surface of the cement grains, mainly characterized by a dot and spatial structure of molecule with branched side chains, which contributes to more effective dispersion of the cement flocculant due to steric effect, as well as to provide access water to clinker minerals. Virtually unlimited varying the amount and length of the side chains allow you to create a controlled plasticizers of adsorption and plasticizing effect depending on the features used cement and aggregates, as well as requirements for the concrete mix. This makes superplasticizers based on polycarboxylate the most promising water-reducing action modifiers and opens wide horizons of their use for high quality cement composites for construction purposes [1–4].

Study of the influence of polycarboxylates on hydration, structure and stability of cement systems is of great importance. Is currently insufficient data on the effect of polycarboxylates on cement systems.

We have studied the impact copolymers of methacrylic acid isobutyl ester (MAAIBE) with sodium salt of acrylic acid (AA) on the properties of cement systems. Copolymers were obtained by block copolymerization in the presence of benzoyl peroxide at various weight ratios of the monomers.

When testing was used cement M400. Spreadability of water–cement mortar was determined according to GOST 26798.1-96. The Effect of superplasticizers on strength of cement stone was measured on samples of size $2 \times 2 \times 2$ cm^3, obtained from the normal consistency of the cement dough, and a control sample at a constant water cement ratio with the addition of plasticizers, hardened samples under normal conditions, and then tested for compressive after 28 days.

As can be seen from Tables 3.1–3.4, a sharp increase in strength of cement by adding plasticizer based on the sodium salt of a copolymer of AA and MAAIBE in range 0.02–0.05% by weight of cement in the

TABLE 3.1 Test Results of Cement Pastes with a Plasticizing Additive Based on a Copolymer of Sodium AA and MAAIBE at a Weight Ratio Percentage: 95:5

№	(Quantity) the amount of cement, g	The amount of additive by weight of cement, %	Water–ce- ment ratio	Spreadability, cm	Strength after 28 days, MPa
1	100	-	0.43	6	17
2	100	0.02	0.40	6	34
3	100	0.05	0.38	6	34
4	100	0.08	0.36	6	30
5	100	0.1	0.35	6	27
6	100	1	0.34	6	27
7	100	1	0.43	8	15

TABLE 3.2 Test Results of Cement Pastes with a Plasticizing Additive Based on a Copolymer of Sodium AA and MAAIBE at a Weight Ratio Percentage: 90:10

№	(Quantity) the amount of cement, g	The amount of additive by weight of cement, %	Water–ce- ment ratio	Spreadability, cm	Strength after 28 days, MPa
1	100	-	0.43	6	17
2	100	0.02	0.41	6	31
3	100	0.05	0.39	6	29
4	100	0.08	0.36	6	28
5	100	0.1	0.35	6	24
6	100	1	0.33	6	20
7	100	1	0.43	12	16

cement–water solution under constant spreadability (decreased w/c ratio), but a further increase in the amount of copolymer decreases the strength of the cement stone.

Spreadability water–cement slurry is increased to 16 cm when adding plasticizer in an amount of 1% of the amount of cement, based on the copolymer of sodium salt AA and MAAIBE in a weight ratio %: 95:5, 90:10, 80:20, but further increasing the amount of monomer MAAIBE the weight ratio to 50%, increases the water requirement of the cement dough.

TABLE 3.3 Test Results of Cement Pastes with a Plasticizing Additive Based on a Copolymer of Sodium AA and MAAIBE at a Weight Ratio Percentage: 80:20

№	(Quantity) the amount of cement, g	The amount of additive by weight of cement, %	Water–cement ratio	Spreadability, cm	Strength after 28 days, MPa
1	100	-	0.43	6	17
2	100	0.02	0.40	6	32
3	100	0.05	0.38	6	33
4	100	0.08	0.36	6	31
5	100	0.1	0.35	6	27
6	100	1	0.34	6	24
7	100	1	0.43	16	15

TABLE 3.4 Test Results of Cement Pastes with a Plasticizing Additive Based on a Copolymer of Sodium AA and MAAIBE at a Weight Ratio Percentage: 50:50

№	(Quantity) the amount of cement, g	The amount of additive by weight of cement, %	Water–cement ratio	Spreadability, cm	Strength after 28 days, MPa
1	100	-	0.43	6	17
2	100	0.02	0.40	6	33
3	100	0.05	0.38	6	34
4	100	0.08	0.36	6	32
5	100	0.1	0.35	6	28
6	100	1	0.48	6	16

3.2   CONCLUSION

Thus, the resulting copolymer based on the sodium salt AA and MAAIBE, when added in different weight ratio is important to investigate the influence of polycarboxylate on the gelation of cement stone. The copolymer dramatically increases the strength of cement stone by adding it in small quantities.

KEYWORDS

- **cement systems**
- **copolymer**
- **methacrylic acid isobutyl ester**
- **Na-acrylic acid**

REFERENCES

1. Gamalii, E. A. Thesis PhD "Complex modifiers based on polycarbohylat ethers and active mineral additions", Chelyabinsk, 2009, 217 pp.
2. Hachnel, C. Interaction Between Cements and Superplasticizers. Proceedings of the 12th International Congress on the Chemistry of Cement. Montreal, 2007, 111–125.
3. Lothenbach, B. The influence of superplasticizers on the hydration of Portland cement. B. Lothenbach, F. Winnefeld, R. Figi. Proceedings of the 12th International Congress on the Chemistry of Cement. Montreal, 2007, 211–233.
4. Pourchet, S. Influence of PC superplasticizers on tricalcium silicate hydration. S. Pourchet, C. Comparet, L. Nicoleau at alias. Proceedings of the 12th International Congress on the Chemistry of Cement. Montreal, 2007, 132–145.

CHAPTER 4

A NOTE ON THE INFLUENCE OF COPOLYMER OF NA-METHACRYLIC ACID AND METHACRYLIC ACID ISOBUTYL ESTER ON THE PHYSIC-MECHANICAL PROPERTIES OF CEMENT SYSTEMS

M. U. KARIMOV,[1] A. T. DJALILOV,[1] and N. A. SAMIGOV[2]

[1]*State Unitary Enterprise Tashkent Research, Institute of Chemical Technology, Tashkent, Uzbekistan*

[2]*Tashkent Architectural and Construction Institute, Tashkent, Uzbekistan*

CONTENTS

ABSTRACT

The chapter explains synthesis and influence of copolymers of Na- methacrylic acid and methacrylic acid isobutyl ester on the physicomechanical properties of cement systems. Cement systems were investigated, according to state standard specifications. Studied the fluidity of water–cement mortar and strength of a cement stone with the addition of chemical additives.

4.1   INTRODUCTION

At the moment there is no clear evidence of an effect of polycarboxylates on the hydration, structure and stability of cement systems. Establishing common patterns of influence of polycarboxylate on the properties of cement compositions is difficult due to the large amount of polycarboxylate additives in the world market [1].

Introduction of polycarboxylates is capable of changing the morphology of the hydrated phases, leading to a decrease in the linear dimensions of the crystals ettringite. There is an opportunity deposition of polycarboxylates for primary hydrosilicates calcium ratio $CaO/SiO_2 \geq 2.5$ and examining the effect on their morphology [2, 3].

Hydration and structure of cement systems, modified polycarboxylates diverge. However, most researchers did not rule out the possibility of slowing down the hydration in the presence of a polycarboxylate plasticizer. In this connection remain unexplained issues associated with stability and the composition formed hydrated phases in the presence of polycarboxylate, respectively, with its resistance and durability [4].

We have studied the impact of copolymers of methacrylic acid isobutyl ester (MAAIBE) with the sodium salt of methacrylic acid (MAA) on the properties of cement systems. Copolymers were obtained by block copolymerization in the presence of benzoyl peroxide at various weight ratios of the monomers.

When testing was used cement M400. Spread ability of water–cement mortar was determined according to GOST 26798.1-96. The Effect of superplasticizers on strength of cement paste was measured on samples of size $2 \times 2 \times 2$ cm^3, obtained from the normal consistency of the cement dough and a control sample at a constant water–cement ratio with the addition of plasticizers, hardened samples under normal conditions and then tested for compressive after 28 days.

As can be seen from Tables 4.1–4.3, a sharp increase in strength of cement paste, adding a plasticizer additive, based on a copolymer of the sodium salt MAA and MAAIBE in small amounts by weight of the cement in the cement–water solution under constant spread ability (reduced w/c ratio), but further increases in the copolymer degrades the strength of the cement stone.

Spread ability of water–cement slurry decreases with increasing monomer IBEMAK in comprising of the copolymer, adding plasticizer in an

TABLE 4.1 Test Results of Cement Pastes with a Plasticizing Additive Based on a Copolymer of Sodium MAA and MAAIBE at a Weight Ratio Percentage: 95:5

№	(Quantity) the amount of cement, g	The amount of additive by weight of cement, %	Water–cement ratio	Spreadability, cm	Strength after 28 days, MPa
1	100	-	0.43	6	17
2	100	0.02	0.40	6	26
3	100	0.05	0.38	6	28
4	100	0.08	0.36	6	30
5	100	0.1	0.35	6	27
6	100	1	0.34	6	26
7	100	1	0.43	14	16

TABLE 4.2 Test Results of Cement Pastes with a Plasticizing Additive Based on a Copolymer of Sodium MAA and MAAIBE at a Weight Ratio Percentage: 70:30

№	(Quantity) the amount of cement, g	The amount of additive by weight of cement, %	Water–cement ratio	Spreadability, cm	Strength after 28 days, MPa
1	100	-	0.43	6	17
2	100	0.02	0.41	6	31
3	100	0.05	0.39	6	29
4	100	0.08	0.36	6	28
5	100	0.1	0.35	6	24
6	100	1	0.33	6	20
7	100	1	0.43	13	16

TABLE 4.3   Test Results of Cement Pastes with a Plasticizing Additive Based on a Copolymer of Sodium MAA and MAAIBE at a Weight Ratio Percentage: 50:50

№	(Quantity) the amount of cement, g	The amount of additive by weight of cement, %	Water–cement ratio	Spreadability, cm	Strength after 28 days, MPa
1	100	-	0.43	6	17
2	100	0.02	0.40	6	33
3	100	0.05	0.38	6	34
4	100	0.08	0.36	6	32
5	100	0.1	0.35	6	28
6	100	1	0.50	6	16

amount of 1% of the amount of cement. Increasing the amount of monomer MAAIBE the weight ratio to 50%, increases the water requirement of the cement paste.

4.2   CONCLUSION

Thus, the resulting copolymer based on the sodium salt MAA and MAAIBE, when added in different weight ratio is important to investigate the influence of polycarboxylate on the gelation of cement stone. The copolymer dramatically increases the strength of cement stone by adding it in small quantities.

KEYWORDS

- cement systems
- copolymer
- methacrylic acid isobutyl ester
- Na-methacrylic acid

REFERENCES

1. Gamalii, E. A. Thesis PhD "Complex modifiers based on polycarbohylat ethers and active mineral additions", Chelyabinsk, 2009, 217 pp.
2. Ledenev, L. A. "Efficiency increasing of the organo-mineral additions application in beton technology." Proceeding of International Congress "Science and Innovations in Building." Voronezh. 2008, 283–287.
3. Hachnel, C. Interaction Between Cements and Superplasticizers. C. Hachnel, H. Lombois-Burgcr, L. Guillot at alias. Proceedings of the 12th International Congress on the Chemistry of Cement. Montreal, 2007, 111–125.
4. Pourchet, S. Influence of PC superplasticizers on tricalcium silicate hydration. S. Pourchet, C. Comparet, L. Nicoleau at alias. Proceedings of the 12th International Congress on the Chemistry of Cement. Montreal, 2007, 132–145.

CHAPTER 5

A NOTE ON SYNTHESIS OF SUPERPLASTICIZER AND ITS INFLUENCE ON THE STRENGTH OF CEMENT PASTE

M. U. KARIMOV and A. T. DJALILOV

State Unitary Enterprise Tashkent Research, Institute of Chemical Technology, Tashkent, Uzbekistan

CONTENTS

ABSTRACT

In this chapter, rheological properties of cement systems and mechanism of action of superplasticizers are studied. This chapter shows the results of analysis of the IR spectra and the resulting impact of the superplasticizer on the strength and the fluidity of cement systems.

5.1 INTRODUCTION

The mechanism of action of superplasticizers is considered by many authors. It is shown that the liquefaction slurry is provided by many factors. Adsorbed on the surface of dispersed particles, molecules of superplasticizer reduce the surface tension at the solid – solution, thereby reducing the tendency of particles to aggregate, with the observed peptization of particles and release the immobilized water, and the increased mobility of the suspensions. In during the plasticization of suspensions are important increase zeta potential of the dispersed phase, the formation of the solvate adsorption layers on the particle surface, the creation of structural and mechanical barrier to the development of other factors aggregation stability [1].

Rheological properties of mineral dispersions are more dependent on the availability and quality of protection layers between the particles. Through these interlayer attraction forces between particles depend on the distance between them, and which are due to van der Waals and hydrogen bonds. As the authors show, the interlayer environment in places of contact, playing the role of lubricant provides mobility of the individual elements of the structure. Hence, by increasing or decreasing the thickness of interlayers of the medium at the contact points of particle or by changing their hydrodynamic properties, applying plasticizers can be adjusted within a wide range coagulation structure of the mechanical properties of the material [2].

We synthesized superplasticizer based on hydrolyzed polyacrylonitrile. Based on indications from the IR spectrum, it can be said that the resulting superplasticiser, mainly has the following functional group. New absorption bands (or spectrums) in the 3346 cm^{-1} show that the functional group – $CONH_2$ changed its structure to – $CONH$ – chemical bond.

In IR spectrum can be shows absorption bands of asymmetric stretching vibrations in the 1150–1260 cm^{-1} and the absorption band characteristic of the symmetric stretching vibrations occur in the region 1010–1080 cm^{-1}. These absorption bands show that the synthesized superplasticizer has a functional group – CH_2SO_3Na.

When testing cement M400 was used. Spreadability of water–cement mortar was determined according to GOST 26798.1-96. The effect of superplasticizers on strength of cement paste was measured on samples of size $2 \times 2 \times 2$ cm^3, obtained from the normal consistency of the cement dough and a control sample at a constant water–cement ratio with the addition of plasticizers, hardened samples under normal conditions and then tested for compressive after 28 days.

As seen from Table 5.1, adding obtained superplasticizer increases fluidity and strength of cement systems at a constant water–cement ratio.

As seen from Table 5.2, with the addition of obtained superplasticizer in an amount of 1% by weight of cement, water/cement ratio decreased from 0.43 to 0.28 and the strength of the cement stone increased to 30 MPa.

TABLE 5.1 Test Results of Cement Pastes with the Addition of Prepared Superplasticizer at Constant W/C Ratio

№	(Quantity) the amount of cement, g	The amount of additive by weight of cement, %	Water–cement ratio	Spreadability, cm	Strength after 28 days, MPa
1	100	-	0.43	6	17
2	100	0.2	0.43	10	20
3	100	0.5	0.43	11	24
4	100	0.8	0.43	14	26
5	100	1	0.43	18	27

TABLE 5.2 Test Results of Cement Pastes with the Addition of Prepared Superplasticizer at Constant Spreadability

№	(Quantity) the amount of cement, g	The amount of additive by weight of cement, %	Water–cement ratio	Spreadability, cm	Strength after 28 days, MPa
1	100	-	0.43	6	17
2	100	0.2	0.40	6	24
3	100	0.5	0.36	6	25
4	100	0.8	0.32	6	27
5	100	1	0.28	6	30

5.2 CONCLUSION

Thus, plasticizers can dramatically changing the rheological properties of disperse systems. Mobility of disperse systems depends on the molecular weight compounds. At a constant water content, more prolonged retention of mobility in time characteristic of the polymers, which, in contrast to the oligomers some change speed setting process.

KEYWORDS

- fluidity
- IR spectra
- strength of cement stone
- superplasticizer

REFERENCES

1. Izotov, V. S., Sokolova, Yu. A. "Chemical Additions for Beton Modification." Moscow: "Poleotip" Publishing House (in Rus.), 2006, p. 244.
2. Ramchadrran, V. S. "Beton addition. Handbook." Moscow: "Stroiizdat" Publishing House (in Rus.). 1988, p. 244.
3. Batrakov, V. G. "Modified Betons." Moscow: "Techoproject" Publishing House (in Rus.). 1998, p. 768.

CHAPTER 6

A NOTE ON INFLUENCE OF NA-CARBOXYMETHYLCELLULOSE ON THE PHYSIC-MECHANICAL PROPERTIES OF CEMENT SYSTEMS

M. U. KARIMOV, G. G. TUKHTAEVA, A. A. KAMBAROVA, and A. T. DJALILOV

State Unitary Enterprise Tashkent Research, Institute of Chemical Technology, Tashkent, Uzbekistan

CONTENTS

ABSTRACT

This chapter reviews the effect of Na-carboxymethylcellulose on physical and mechanical properties of on cement systems. The main goal is to review the effect of the resulting plasticizing additive for strength and fluidity of cement systems. Compared the fluidity of water–cement mortar by adding obtained Na-carboxymethylcellulose with other Na-carboxymethylcellulose.

6.1   INTRODUCTION

Plasticizers are the most widely used in the production of concrete and reinforced concrete, due to their high efficiency, the lack of a negative effect on the properties of concrete and condition of reinforcement.

The main purpose in this field used plasticizers – increasing the mobility of the concrete mix, which is important to facilitate the molding process of designs, increasing the density and strength of the concrete by reducing the water requirement of the concrete mix, while maintaining its original mobility, or to reduce the consumption of cement [1].

According to modern concepts plasticizers – surface-active substance (SAS) are dispersants – stabilizers, as a result of adsorption on the surface of solid and liquid phases, structured film. Immobilization bound in flocculation cement water, reducing the internal friction coefficient of cement–water suspension, in some cases, increasing the electrostatic repulsion of the particles due to a significant increase in their electrokinetic potential – the main factors that affect the plasticizing effect of SAS on the cement–water systems to reduce their water demand and consumption of binder.

Plasticizing SAS significantly improve the fluidity of cement paste by reducing the surface tension at the interface. This greatly reduces the water requirement of the concrete mix. Additives of this group are the most effective plasticizers in concrete mixtures at relatively high flow cement [2].

As the plasticizer used in concrete mixtures also cellulose derivatives such as methylcellulose, carboxymethylcellulose, etc.

The degree of polymerization (DP) Na-carboxymethylcellulose is important when using it as a plasticizer. Use of concentrated solutions of alkali during mercerization of cellulose in turn affects on DP Na – carboxymethylcellulose. We have obtained Na – carboxymethylcellulose based linters using low concentrated solutions of alkali during mercerization. Figure 6.1 shows the results of studying the spreadability of water–cement mortar with the addition of Na – carboxymethyl cellulose, obtained on the basis of domestic cellulose (Na-CMC – 1), foreign Na – carboxymethyl cellulose (Na-CMC – 2) and Na – carboxymethyl cellulose, obtained by us using lint (Na-CMC – 3).

As can be seen from the graph, the effect of the obtained Na-carboxymethyl cellulose on the fluidity of water–cement mortar significantly. With

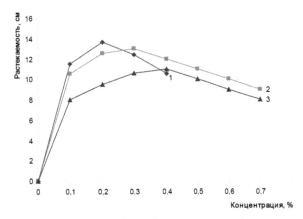

FIGURE 6.1 Dependence spreadability of water – cement slurry to concentration: 1 – Na-CMC – 3; 2-Na-CMC – 2; 3 – Na-CMC – 1.

further increase in the concentration of all samples Na-carboxymethylcellulose reduced fluidity of water – cement mortar.

As seen from Table 6.1, with the addition of obtained Na-CMC, when small amounts, increases fluidity sharply, but the further increase additives lead to increased water demand of cement paste. Strength of cement increased by adding the resulting Na-CMC in the range of 0.2–0.8% by weight of cement and a further increase the addition to 1% is not observed a sharp change in the strength of cement stone. With further study, adding the obtained Na-CMC above 1%, led to a deterioration in strength of cement stone.

TABLE 6.1 Test Results of Cement Pastes with the Addition Obtained Na-CMC – 3

№	(Quantity) the amount of cement, g	The amount of additive by weight of cement, %	Water–cement ratio	Spreadability, cm	Strength after 28 days, MPa
1	100	-	0.43	6	17
2	100	0.2	0.43	14	22
3	100	0.5	0.43	12	26
4	100	0.8	0.43	10	26
5	100	1	0.48	8	27

6.2   CONCLUSION

Thus, fluidity of cement- water mortar with adding the obtained Na-CMC, with small amounts, increases in comparison with other Na-CMC. Degree of polymerization of Na-CMC plays an important role not only in the manufacture of cement systems, but also in other areas of the economy where Na-CMC applied.

KEYWORDS

- cement system
- degree of polymerization
- mercerization
- Na-carboxymethylcellulose
- spreadability

REFERENCES

1. Izotov, V. S., Sokolova, Yu. A. "Chemical Additions for Beton Modification. " Moscow: "Poleotip" Publishing House (in Rus.), 2006, 244.
2. Ibragimpov, R. A. Thesis PhD "The Hard Betons with Complex Addition of Polycarbohylat Ethers." Kazan. 2011, 184.

CHAPTER 7

A NOTE ON CHANGE OF MORPHOLOGICAL FEATURES CELLULOSE IN DIFFERENT ACID-CATALYTIC SYSTEMS

KUVSHINOVA LARISA ALEXANDROVNA[1] and
MANAHOVA TAT'YANA NICOLAEVNA[2]

[1]*Federal State Institution of Science, Institute Chemistry of Komi Scientific Centre of the Ural Branch of the Russian Academy of Sciences, Pervomaiskaya St., 48, Syktyvkar, 167982, Russia, E-mail: fragl74@mail.ru*

[2]*Northern (Arctic) Federal University named after M.V. Lomonosov, Northern Dvina Emb., 17, Arkhangelsk, 163002, Russia, E-mail: tatiankaya17@yandex.ru*

CONTENTS

ABSTRACT

This chapter studies the effect of various acid-catalytic systems on the degradation of unbleached softwood cellulose. Evidence on the extent of change cellulose fibers as a result of chemical action was obtained by the device L&W fiber tester. Shown, that solutions of titanium tetrachloride in hexane as compared to others acid-catalytic systems are most effective for degradation of macromolecules cellulose.

7.1 INTRODUCTION

Application of acid-catalytic systems for the degradation cellulosic raw material was directed on obtaining powdered products, demanded in various fields of science and technology. Most often, for this purpose are used a heterogeneous hydrolytic treatment in solutions Bronsted mineral acids at reflux.

Solutions Lewis acids to nonpolar organic solvents, which also are leading to chemical degradation of various semis cellulose (from herbaceous and from woody vegetable raw material) [1], is alternative to the aforementioned acid-catalytic system. Among the Lewis acids, which are halide of elements variable valency, favorably differs titanium tetrachloride ($TiCl_4$) in hexane (C_6H_{14}) by effective action of degradation cellulosic glycosidic bonds. This system is not yet applied for the production of cellulose powder on an industrial scale. However, its use for the processing of cellulosic raw material to such products is very promising. The possibility of multiple simple regeneration solvent and reducing the amount of waste water (or the lack thereof), and also decrease in power inputs (no longer needed to maintain a high temperature and pressure for the reaction) all these are does attractive for use of solutions $TiCl_4$ in C_6H_{14}.

The aim of this study is a comparative evaluation of the effectiveness of destructive action of such the acid-catalyst systems as titanium tetrachloride in hexane and hydrochloric acid in hexane and in water on the cellulose fibers according changes the morphological features of the latters.

To prepare the acid-catalyst systems used commercial $TiCl_4$, and C_6H_{14} of production "Vekton" qualification "clean," which were purified before using by simple distillation [2]. Before using C_6H_{14} in it got rid from a small amounts of water by dehumidification metal alloy of sodium with potassium at reflux under argon (indicator – benzophenone). A concentrated hydrochloric acid of qualification "analytically pure" manufactured by JSC "Caustic" (Republic of Bashkortostan, Russia) was used for this work also.

The unbleached cellulose from softwood manufactured by JSC "Mondi Syktyvkar Forest Industry Complex" (Komi Republic, Russia) was used as object of researching. Average value of degree polymerization (DP_{av}) of the original cellulose was 1260 units. The cellulosic fibers were subjected to drying to constant weight in an oven at $103 \pm 2°C$ for removal of physically bound water. Then treatment of cellulose was carried out in different acid-catalyst systems at $22 \pm 1°C$ and at relation of liquid to solid 20 cm^3: 1 g (liquid module), using intensive mixing the cellulosic suspension. The treatment conditions are shown in Table 7.1.

According to the results shown in Table 7.1, the above acid- catalyst systems under the same conditions of application had different effectiveness of destructive action on cellulosic fibers. It is known that the system $HCl – H_2O$ in conditions of treatment 2.5 M. solution at $105 \pm 1°C$ is

TABLE 7.1 Influence the Conditions of Treatment of Cellulose on the Characteristics Products of Its Degradation

№	Conditions of treatment			Characteristics products		
	Acid-catalyst systems: reagent – medium	$C_{reagent}$, mol/dm^3	DP_{av}	Average size		
				l, mm	d, mcm	
1	Without treatment	-	1260	2.20	28.8	
2	$HCl – H_2O$	17.5×10^{-3}	1210	2.18	27.9	
3	$HCl – H_2O$	2.5	1060	2.12	27.7	
4	$HCl – C_6H_{14}$	17.5×10^{-3}	1190	2.12	28.1	
5	$HCl – C_6H_{14}$	2.5	220	0.34	31.2	
6	$TiCl_4 – C_6H_{14}$	17.5×10^{-3}	470	0.85	29.6	
7	$*TiCl_4 – C_6H_{14}$	17.5×10^{-3}	190	0.29	33.5	

*Designations: l and d – the average sizes of length and diameter of cellulosic fiber, respectively; * – sample 7 was prepared using air-dry cellulosic fibers without a preliminary drying to constant weight, the duration of treatment was 15 min instead of 210 min as in the remaining samples 2 ÷ 6.*

"classical" for the production cellulosic powder [3]. Although application this system at lower temperature is not significantly effect to reducing the value DP_{av} of cellulose and obtaining free flowing products. Increasing the concentration of solution HCl in H_2O from 17.5×10^{-3} mol/dm^3 to 2.5 mol/dm^3 at treatment cellulose resulted to reduce value DP_{av} of cellulose on 4 and 16% respectively. Replacing in system H_2O on C_6H_{14} in the presence of the maximum amount of the reagent gave better the result of degradation, reducing the value of DP_{av} by more than 80%.

Known, that Lewis acids such as $TiCl_4$ are easily hydrolyzed in the presence of insignificant amounts of H_2O in the system. Therefore, for comparison of results degradation the treatment cellulose in system $TiCl_4 - C_6H_{14}$ was carried before and after removing the possible amounts of physically bound water from fibers. Use of a Lewis acid solution of low concentration lets to obtain shorter fibers cellulose (sample 6) and cellulose powder (sample 7) with values $DP_{av,}$ which equal 470 and 190, respectively. In these samples compared to the others, obtained in the above-mentioned acid-catalyst systems, values DP_{av} managed to reduce by treatment on 60 and 85%. Cellulose that not was subjected to drying before treatment lends itself to faster destruction in system $TiCl_4 - C_6H_{14}$. The duration of the treatment in the latter case was 15 min, which is 14 times less than in the other experiments.

From Table 7.1 shows that characteristics of degradation products such as value DP_{av} cellulose and size l fibers directly proportional are inter-related. The average square deviation of correlation according to the equation of the line: $l = 0.55 \times DP_{av}$ equal 0.99. The obtained data let to determine the unknown characteristic on the basis of known characteristic in the equation (among products derived by degradation of mentioned unbleached cellulose softwood).

The average diameter of cellulose fibers treated in different acid-catalyst systems changes also. As the result use of solutions HCl in H_2O and in C_6H_{14} the size d is reduced (samples $2 \div 4$, Table 7.1). Water used at the treatment and at the follow washing of samples leads to the dis-solution and to the washing out water-soluble fraction of the particles with low molecular weight, which is formed in the result destruction cellulose. The sample 5 is characterized of higher value d as compared with such value of the original cellulose. This may be associated with feature of the system $HCl - C_6H_{14}$, which is a two-phase immiscible liquid. Reaction of cellulose destruction, caused by the action of HCl, largely took place in the lower layer. And since the amount of this reagent in the system was significantly higher (sample 5) than in the previous experiment (sample 4), it could promote greater swelling [4], and loosening of the cellulose fibers in concentrated HCl. In the result size d is increased.

Samples 6 and 7 obtained in the system $TiCl_4 - C_6H_{14}$, also have an increased size d by adsorption on the surface compounds of transformed Lewis acid [5], from which if necessary it is easy to release.

For samples $1 \div 7$ were obtained a results of the fractional distribution of cellulose fibers along their length (shown in Figures 7.1 and 7.2) by an optical device L&W Fiber Tester.

According to Figure 7.1 fractional distribution of length cellulose fibers in samples $2 \div 4$ is almost identical to that in softwood unbleached cellulose before treatment (sample 1) and has a kind of bimodal curve with two maxima in the length range of 0.5–1.0 mm and 2.5 – 3.0 mm.

The sample 6 gained form of short-fiber in the result of cellulose degradation therefore it was analyzed by the above-mentioned device at two different conditions, set as for long-fiber samples $1 \div 4$, so and for samples 5 and 7 in powdered form. In the first case (Figure 7.1) sample 6 differs more narrow distribution of fiber length with a maximum in the region of 0.5–1.0 mm as compared with long-fiber samples. In the second case (Figure 7.2), this maximum is located in the total amount of length ranges, which are corresponding to 0.60–7.50 mm. It does not contradict to the results given in the previous Figure 7.1.

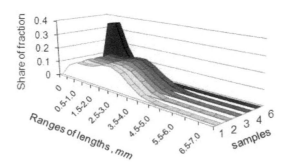

FIGURE 7.1 Comparative distribution of cellulosic fiber length in samples $1 \div 4$ and 6 (numbering of samples on the Figure 7.1 and in the Table 7.1 is corresponds).

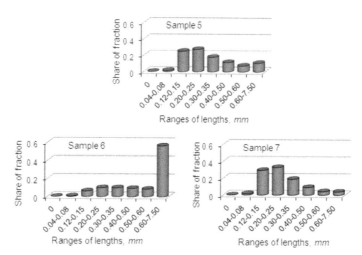

FIGURE 7.2   Fractional distribution of length cellulose fibers in samples 5 ÷ 7 (numbering of samples on the Figure 7.2 and in the Table 7.1 is corresponds).

Fractional distribution of length cellulose fibers in the sample 7 (Figure 7.2) is most narrow and has a maximum at 0.20–0.25 mm. The specified length range is the share equal 0.33 from all factions of this sample. Sample 5 (at comparison to the sample 7) differs less marked and more gradual transition from one fraction to another. It is also at a relatively low value DP_{av} has a maximum in range length of 0.20–0.25 mm, equal 0.27.

7.2   CONCLUSIONS

On example the unbleached cellulose from softwood was confirmed that the solution of titanium tetrachloride in hexane has a strong destructive effect on fibers of cellulose. The resulting powdered product is characterized by the narrowest distribution of cellulose fibers on length. This alternative method of degradation of cellulose is allows to receive on the world market a competitive analogs of cellulose powder in less energy-intensive conditions of treatment.

ACKNOWLEDGEMENTS

The authors gratefully acknowledge the partial financial support of the Project 12-P-3–1024 "Adapting the method of pK-spectroscopy to the study of acid-base properties and structure of ionizable biopolymers and their derivatives" (the program of the Presidium of the Russian Academy of Sciences).

Working on an automatic analyzer fiber L&W Fiber Tester was made in innovation and technology center "Modern technologies for processing biological resources of the North" (Northern (Arctic) Federal University named M.V. Lomonosov) at financial support from the Ministry of Education of Russia."

KEYWORDS

- **fractionation cellulose fibers**
- **length and diameter of the cellulose fiber**
- **titanium tetrachloride**
- **unbleached softwood cellulose**

REFERENCES

1. Frolova, S. V., Kuvshinova, L. A., Kutchin, A. V. Method Obtaining of Powder Cellulose. Patent of the Russian Federation N 2478664. Registered 10.04.2013.
2. Karjakin, Y. V., Angelov, I. I. Pure Chemical Substances. Moscow: "Khimiya" ("Chemistry", in Rus.) Publishing House, 1974, 408.
3. Kocheva, L. S., Karmanov, A. P. New methods of producing microcrystalline cellulose. Chemistry and Technology of Plant Substances: Abstracts of II Russian Conference. Kazan, 2002, 140.
4. Varviker, D. O. "Swelling Cellulose and Its Derivatives" Ed. N. Bayklz and L. Segal. Moscow: "Mir" (Peace" in Rus.) Publishing House, 1974, Vol. 1, 235–279.
5. Kuvshinova, L. A., Manahova, T. N. Influence Solution of Titanium Tetrachloride on Morphological Features of Cellulose. Chemistry of Plant Raw Materials. In edition, registration number of 130729/01.

CHAPTER 8

A NOTE ON SYNTHESIS AND PROPERTIES OF PBT

M. M. LIGIDOVA, T. A. BORUKAEV, and A. K. MIKITAEV

Kabardino-Balkarian State University 173, Chernyshevsky Str., 360004, Nalchik, Russia

CONTENTS

ABSTRACT

In this chapter, an experimental research on synthesis, properties study and application of polybutylene terephthalate is carried out (PBT). To modify the properties in PBT are introduced (in an amount of 2–80%) such fillers as glass fiber, carbon fiber, chalk, talc, graphite, and flame retardants and other polymers.

8.1   INTRODUCTION

One of the most important representatives of industrial polyesters is polybutylene terephthalate (PBT) found wide application in engineering, aircraft industry, electrical engineering, medicine, and recently in the production of fiber optic cables. PBT is a successful combination of technological and operational properties [1].

The process of crystallization of this polymer occurs very quickly, so the cycle time is reduced to a minimum. The PBT has excellent weather resistance, resistance to wear, dimensional stability and balanced physical properties. The properties of the polymer are greatly enhanced, if the material is filled with fiberglass or mineral supplements [2].

Modification of PBT by introduction into it fiberglass, mineral fillers and modifying additives can significantly change the properties of the basic material and impart to the compositions on its basis:

- higher heat resistance;
- significant increase its rigidity and durability;
- to give the basic material flame-proof properties, etc.

Temperature of long use of PBT without changing the mechanical and dielectric properties is 120°C, for glass-filled PBT up to 155°C, short-term to 210°C.

Polybutylene terephthalate soluble in mixtures of phenol with chlorinated aliphatic hydrocarbons, in m-Cresol. Insoluble in aliphatic and perkhlorirovannykh hydrocarbons, alcohols, ethers, fats, vegetable and mineral oils and various types of motor fuel. At 60°C it is not guide stable limited racks in diluted acids and alkalis, but destroyed in concentrated mineral acids and alkalis. By their resistance to the action of the chemical reagents and cracking it exceeds polycarbonates.

To modify the properties (in of quantity 2–80%) fillers (glass fiber, carbon fiber, chalk, $BaSO_4$, talc, graphite and other), flame retardants (brominated organic substances in combination with Sb_2O_3), polymers (polyethylene terephthalate), polycarbonates, titanium dioxide), dyes are introduced into PBT.

This polyester is mainly processed by injection molding, less seldom by. An important advantage of this substance before other thermoplastics (work with polyethylene terephthalate, polycarbonate, polysulfonate) – good technological properties associated with a high rate of crystallization at low temperatures (30–1000°C) and high fluidity of melt.

Casting PBT and composite materials on its basis are substituted by metals (zinc, bronze, aluminum) and thermosets in the production of electrical parts (high-voltage parts of ignition systems, plugs and sockets, brush holders, housing relay coils etc.), construction (e.g., housing, clips, gears, bearings) and the decorative (details of decoration, blinds and others) assignments in the automotive, electrical, electronics, household appliances.

Extruded polybutylene terephthalate is used for production of films, rods, tubes, profiles, fibers.

8.2   EXPERIMENTAL PART

8.2.1   GETTING POLYBUTYLENE TEREPHTHALATE

The given polyester is obtained in two stages on periodic or continuous schemes. At the first stage bis-(4-gidroksibutil)terephthalate, is synthesized at the second one spend polycondensation is carried ant.

The stage of process is an equilibrium reactions of nucleophilic substitution at carbon atom of carbonyl group and it can catalyzed by both weak acids and by bases. More often as catalysts various salts (it's acetates, carbonates, borates, phosphates, chlorides and oxides of antimony, tin, manganese, aluminums, titanium, zinc, lead, sodium, calcium, mercury, iron, copper, etc., are used.

Polycondensation of bis (4-gidroksibytil)tereftalat carried out in a vacuum at 240–250°C; cat.-Ti(OC$_4$H$_9$)$_4$. Melt polybutylene terephthalate is being ousted from the autoclave, cooled and crushed to the cylindrical pellets. Granulate is dried in vacuum or air driers [2] (Table 8.1).

TABLE 8.1 Physical and Chemical Properties of PBT Compared to Other Engineering Plastics

Indicator	Pet tere-phthalate	PTTF*	PBT	PA 6	PA 6,6
			Material		
Temperature of melting, (°C)	260	228	224	220	265
The glass transition temperature, (°C)	70–80	45–55	20–40	40–80	50–90
Density (amorphous polymer), (g/cm³)	1.335	1.277	1.286	1.11	1.09
Density (crystalline polymer), (g/cm³)	1.455	1.387	1.39	1.23	1.24
Index crystallization rate	1	10	15	5	12

8.3   RESULTS AND DISCUSSIONS

As distinct from polyethylene terephthalate, polybutylene terephthalate is a quickly crystallized polymer; the maximum degree of crystallinity is 60%. It has high strength, stiffness and hardness, resistant to creep, find a good insulator.

The polyester has good sliding properties. Coefficient of friction of polybutylene terephthalate is significantly less than that of poly e-caproamide and polyformaldehyde [3].

Unlike polyamides, polybutylene terephthalate, thanks to minor water absorption, are high dielectric and mechanical properties stored in conditions of increased humidity. At prolonged contact with water and water solutions of salts (e.g., $NaHCO_3$, Na_2CO_3, $NaHSO_3$, KCl) polyester is undergone hydrolytic destruction: the rate of the process at room temperature is negligible, but it increases at higher temperatures (from 80°C).

8.4   CONCLUSION

The last two decades witnessed the intensive development of global production PBT. By estimations of experts the world market volume PBT in 2005 can be estimated at around 700 tons, and the annual average growth rate of consumption is about 7%. The sharp increase in the need for PBT on the international market forced the main suppliers of this polymer to increase the production capacity for is receiving [4].

KEYWORDS

- **additives**
- **fillers**
- **flame retardants**
- **polybutylene terephthalate (PBT)**
- **polycondensation**
- **properties**
- **resin**
- **synthesis**

REFERENCES

1. Patent Russia №2052473, MKI C 08 L 67/02, 1992.
2. http: //www.maltima.ru.
3. Petukhov, B. C., Polyester Fibers, M., 1976; Wilkes, G. L. Sladowski, E. L., "Journal Applied Polymer Science." 1978, 22, 766–769.
4. Margolis, J. M., Engineering Thermoplastics: Properties and Applications, NY, 1985, 19–27.
5. http: //www.newchemistry.ru/inde.

CHAPTER 9

A NOTE ON EXPERIMENTAL DATA BY PIECEWISE LINEAR FUNCTIONS

OSHKHUNOV MUAED MUZAFAROVICH and DZHANKULAEVA
MADINA AMERHANOVNA

*Kabardino-Balkarian State University of H.M. Berbekov, Moscow,
Russia*

CONTENTS

ABSTRACT

The effective data processing algorithm is offered, which is the approximation of the plots of experimental points on the plane by piecewise linear functions. The idea of the method in the choice of the parameters of a linear function on a minimum sum of squared distances. Unlike the classical least squares method, this approach reduces the dispersion and leads to the dual selection of the parameters of the linear function. This algorithm is implemented as a program, the numerical experiments showed its high effectiveness.

9.1   INTRODUCTION

In the study of the properties of polymers is necessary to scientifically based analysis of experimental data in order to describe the statistical regularity as possible. In other words, should approximate the whole set of experimental data by piecewise linear functions so that the dispersion (or deviation) of unknown function in the experimental data was minimal. To solve this problem, usually the least squares method is used in which minimize the sum of squared deviations of the axis Oy of the plane. The difference of this approach from the classic is minimizing the sum of squares of the distances from these points to a linear function, which leads to a decrease of mean square deviation non-uniqueness and of select line.

For explaining these ideas, consider the following problem. Required to find a piecewise linear function, which is optimal for groups of experimental points (*see* Figure 9.1).

Here through $\left(\overline{x}_i, \overline{y}_i\right)$ designated coordinates plots boundaries, and the experimental results themselves are presented as asterisks coordinates $\left(x_i, y_i\right)$.

Assume that the initial point of the experiment is $\left(\overline{x}_0, \overline{y}_0\right)$. Then the choice of line direction or slope is determined by solving the equation (in the case where $\sum \tilde{x}_i \tilde{y}_i \neq 0$)

$$k^2 - \alpha k - 1 = 0 \tag{1}$$

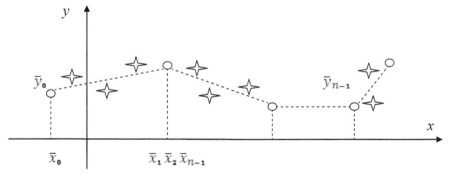

FIGURE 9.1   Scheme for constructing the graph plots.

where

$$\alpha = \frac{\sum_{i=1}^{n_1}\left(\tilde{x}_i^2 - \tilde{y}_i^2\right)}{\sum_{i=1}^{n_1}\tilde{x}_i^2 \tilde{y}_i^2}$$

$$\tilde{x}_i = x_i - \overline{x}_0, \; \tilde{y}_i = y_i - \overline{y}_0$$

$$\overline{x}_0 = \left(\sum_{i=1}^{n_1} x_i\right)\Big/ n_1, \; \overline{y}_0 = \left(\sum_{i=1}^{n_1} y_i\right)\Big/ n_1 \tag{2}$$

Equation (1) is derived by minimizing the sum of squared distances from the points with coordinates (x_i, y_i), $i = 1 \div n_1$ to line $y = kx + b$, that is, from a minimum of expression

$$S = \frac{\sum_{i=1}^{n_1}\left(kx_i + b - y_i\right)^2}{k^2 + 1} \tag{3}$$

Here n_1 – the number of points (asterisks in Figure 9.1) to the point with the abscissa $\overline{x}_1$, i.e., $x_i \le \overline{x}_1$.

Equation (1) at condition $\sum_{i=1}^{n_1}\tilde{x}_i\tilde{y}_i \ne 0$ obviously has two solutions

$$k_{1,2} = \frac{\alpha \pm \sqrt{\alpha^2 + 4}}{2} \tag{4}$$

Corresponding to these solutions linear sections of graphics function have the form:

$$y = k_1 x + b_1, \quad y = k_2 x + b_2,$$
$$b_1 = \overline{y}_0 - k_1 \overline{x}_0, \quad b_2 = \overline{y}_0 - k_2 \overline{x}_0 \tag{5}$$

From the form of Eq. (3) that the lines (5) perpendicular to each other and pass through the point with coordinates $(\overline{x}_0, \overline{y}_0)$.

Question of choosing one of the two optimal lines can be solved by minimizing the sum of squared distances (3); if

$$S(k_1, b_1) < S(k_2, b_2) \tag{6}$$

then we can choose the linear portion of the form $y = k_1 x + b_1$ and on the contrary.

Result of realization of this approach in MatLab presented in Figure 9.2. Here graphics consists of two sections. For the first section we obtain the equations of two lines (according to Eq. (5)), then the Eq. (3) calculate the sums $S(k_1, b_1)$ and $S(k_2, b_2)$. Choice of the optimal line is carried according to condition (6). Then proceed to find the optimal line for the second portion by a similar algorithm.

It is easy to prove that mean square deviation when choosing a piece-wise linear function of the minimum of the Eq. (3) is less than the classic version in $\sqrt{k^2 + 1}$ times. This means that the use of a new algorithm (not classical) more substantiated at large values of parameter k. In the case when a statistical dependence between the two values is weak, that is, then $|k| \approx 0$, both approaches lead to almost identical results.

In Ref. [3], the application of this method for obtaining analytical dependences between stresses and strains for mixtures of PVC+SVN-40 with different ratios of the components are considered.

It should be noted that this algorithm is also suitable for solving such practical importance problems, as tracing roads, which the minimum distance from human settlements, optimal design of water and gas networks, etc.

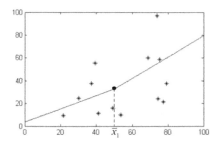

FIGURE 9.2   Choosing to lines and minimizes the sum of squared distances.

KEYWORDS

- **data processing algorithm**
- **dispersion**
- **method of least squares**
- **piecewise linear approximation**

REFERENCES

1. Oshkhunov, M. M., Pisanova I.S. Placement of investments under uncertainty. Proceedings KBSC RAS (in Rus.), 2006, № 2 (15).
2. Zhazaeva, E. M., Thakahov, R. B., Oshkhunov, M. M. Strength and fracture energy of heat-treated polymer blends based on PVC and SKN. Plastic Masses. Moscow: "Khimiya" ("Chemistry", in Rus.) Publishing House. 1975, 54–57.

PART II

COMPOSITES TECHNOLOGY

CHAPTER 10

COMPOSITE ELECTROCONDUCTIVE MATERIALS BASED ON POLYANILINE

S. G. KISELEVA, A. V. ORLOV, and G. P. KARPACHEVA

TIPS RAS, Moscow, Russia

CONTENTS

ABSTRACT

The effect of substituted aromatic amines on the process of heterophase oxidative polymerization of aniline and the quality of the formed polyaniline layers was studied for regulation of characteristics of composites based on them. It was shown that p-phenylenediamine (p-PDA), being a catalyst of the oxidation process at the start of the reaction produces practically no effect on the rate of polyaniline formation at the interface. M-phenylenediamine (m-PDA) shows strong inhibitory properties by increasing the induction period of the reaction and decreasing the rate of the reaction. Diphenylamine (DPA) produces practically no effect on the kinetic parameters of the process, but it impairs the quality of the forming polyaniline layers. Composite films, formed on the basis of polystyrene solution and polyaniline microlayers, obtained directly during the reaction, are characterized by homogeneity and excellent electrophysical and strength properties.

10.1 INTRODUCTION

Polyaniline is one of the most studied representatives of the class of electro-active polymers. It takes a special place among the polymers with a system of polyconjugation due to the elegance of synthesis, ease of the conduction of reversible doping processes – dedoping and valuable electrophysical characteristics. However, the stiffness of the polymer chain and, as a consequence, infusibility and insolubility in most organic solvents, and also insufficient stability in the operating conditions are serious barriers for its practical application. The range of organic solvents of polyaniline is quite limited, and the obtained films and coatings have poor mechanical properties and to improve them one has to introduce additives, for example polystyrene or polyvinylchloride or to obtain complexes with sulfonic acids. In this case the procedure for obtaining films and coatings is complicated and multistage. That is why the problem of creating a technology for obtaining quality films and coatings based on polyaniline, for application in organic electronics, in electrochromic systems, as protective shields from electromagnetic and radioactive radiation etc., is still actual nowadays. One of the ways to solve this problem is the formation of composite materials, formed directly during the reaction of the interfacial polymerization, it allows to simplify the obtaining of new materials based on polyaniline.

This method of obtaining polymer-polymer compositions is based on the reaction of interfacial polymerization of aniline, which was conducted for the first time in our laboratory [1–3]. This reaction allows to obtain nanolayers of polyaniline on the surface of substrates of different chemical nature and in different states of aggregation (both as a solid and as a liquid) directly during the synthesis.

10.2 RESULTS AND DISCUSSION

It was found that polymerization of aniline in heterophase conditions on the interface of two immiscible liquids occurs with a formation of thin polyaniline films with a thickness of 0.2–2 mcm, depending on the reaction conditions. Thus there is an encapsulation of particles of the dispersed phase with the formation of polyaniline microspheres with a diameter up to 70–100 mcm (Figure 10.1a). It was shown that the biggest effect on the course of the polymerization reaction is made by the ratio of reagents concentration and the size of interfacial surface. The best results in uniformity of polyaniline films and minimizing processes of oxidative polymerization in aqueous medium were obtained at the concentration of reagents 0.02 mol/L and the ratio of aqueous and organic phases 1/1. In this case the organic solvent acts as a dispersed phase. The use of polystyrene solution in benzene (3–5 wt.%) as a dispersed phase does not lead to any significant changes in both kinetic parameters of the reaction and the chemical structure of the formed polyaniline films.

The further increase of polymer concentration leads to the increase of the size of microcapsules up to 150 – 300 mcm (Figure 10.1b) and to some slowing of the heterophase polymerization. It is described by the increase of viscosity of the organic phase and, as a consequence, decrease of its dispersion, which affects the rate of the interfacial processes. According to IR-spectroscopy, the obtained polyaniline practically does not differ from the polymer, obtained via polymerization in homogeneous aqueous medium, while the part of the defective units decreases.

However, varying the concentration of reagents and content of the polymer do not significantly change the quality characteristics (chemical

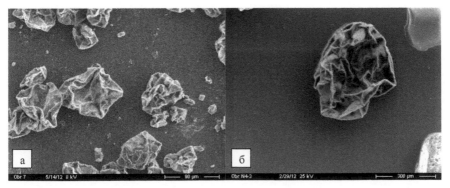

FIGURE 10.1   Electron microscopic image of polyaniline capsules, formed in heterophase conditions: (a) 2.5% solution of PS, (b) 5% solution of PS.

structure, morphology, electrophysical properties) of the forming polyaniline particles and practically does not effect on process control.

The effect of various additives that can largely determine the rate of the oxidation reaction and the value and structure of the interfacial surface, which finally gives an opportunity to control the quality of the forming polyaniline layers and heterophase process as a whole, was studied. Comonomers of aniline catalyzing (p-PDA) and inhibiting (m-PDA) the process of oxidative polymerization by changing the oxidative potential were investigated as active compounds. It was found that introducing into the reaction system up to 5 wt.% p-PDA in relation to the monomer leads to a dramatic reduction of induction period of oxidative polymerization from 18 to 2 min and an insignificant growth of process rate as a whole, as it can be seen from Figure 10.2 (curve 2). It tells about catalytic activity of p-PDA on the first stage of the reaction and can be described by its lower oxidation potential as compared to aniline. At the same time it acts as an active mediator of one-electron oxidation of aniline via coordination with monomer, forming complexes of the semiquinone type. During the formation of polyaniline structures they themselves act as catalysts of one-electron oxidation of the monomer, because their oxidation potential is significantly lower then not only of the monomer, but also of the p-PDA. Moreover, the coordination of the monomer with polyaniline is more effective then with p-PDA. Thus, on the first stage of the oxidation the catalysis of p-PDA proceeds, which is reflected in significant reduction of time of the induction period. In the autocatalytic process, developing further, p-PDA does not participate, which explains the slight growth of polymerization rate in its presence.

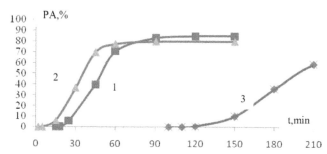

FIGURE 10.2 Dependence of the polyaniline on time: (1) in the presence of additives; (2) in the presence of 5% p-PDA; (3) in the presence of 2% m-PDA.

According to IR-spectroscopy data, an increased content of end amino groups and formations corresponding to 1–2 joining is observed in samples, obtained in presence of p-PDA. It shows that there is some loss of molecular mass of polyaniline and indicates the presence of defect structures. All these facts point out that p-PDA integrates into the growing polyaniline chain.

Addition of p-PDA to the reaction mixture changes not only kinetic parameters of heterophase polymerization, but also leads to the morphological changes of polyaniline films. This results in reducing of the size of capsules, forming during the interfacial polymerization from 100 to 50 mcm (Figure 10.3a). Since the rate of growth of polyaniline films at the start of the reaction on the surface of microdroplets of PS solution in presence of p-PDA is much higher, so it prevents their destruction and coalescence during polymerization.

m-PDA has an inhibitory effect on the heterophase oxidative polymerization of aniline, which reflects in significant growth of induction period. Introduction of 2wt.% m-PDA causes an increase of induction period from 20 min to 2 h (Figure 10.2, curve 3) and a significant reduction of reaction rate. During this process there occurs a formation of capsules of much bigger size (200–300 mcm), which easily break down into separate particles with a size of 1–2 mcm (Figure 10.3b).

Almost without participating in catalysis, m-PDA reacts with aniline and the growing polymer chain, forming end groups of m-substituted aromatic structures type as a result of recombination reaction, which impede

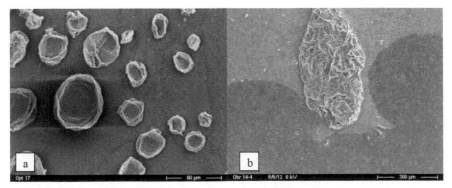

FIGURE 10.3 Electron microscopic image of polyaniline capsules, forming in heterophase conditions at the adding to the reaction mixture: (a) 5% p-PDA, (b) 2% m-PDA.

further oxidation and growth of the polymer chain. It leads to a sharp reduction of reaction rate and formation of low molecular weight products, decreasing the entire yield of polyaniline. According to IR spectroscopy, there is a significant part of phenazine fragments in the obtained polymers, which are a result of m-PDA introduction into the polymer chain and further reaction of oxidative cyclization.

N-phenyl substituted aniline – diphenylamine (DPA) is able to participate in interfacial oxidative polymerization of aniline. After addition of 1–5 wt.% with respect to the monomer DPA to the solution of polystyrene in benzene, its copolymerization with aniline in heterophase conditions take place. Kinetic studies showed that DPA does not affect the course of the oxidation reaction. It shows that the primary processes of monomer oxidation proceed in aqueous solution without DPA participation. However, according to IR spectroscopy data, the obtained polyaniline contains diphenyl fragments. Low molecular oxidated fragments, forming at the start of the reaction, probably have a greater affinity for the organic solvent and pass to it from the interface, where DPA participates in oxidation and addition to the growing chain. Since the end DPA fragments do not prevent the further growth of the polyaniline chain, its presence does not affect the overall kinetics of the process. A slight decrease of molecular mass and an increase of solubility in organic solvents occur.

There is a significant degradation of polyaniline film quality in presence of DPA, as it is seen from Figure 10.4. This, apparently, can be explained by the effect of DPA on the structure of the double electric layer (DEL), forming at the start of the reaction. As was shown earlier [3], it is the DEL, which determines the growth of polyaniline film at the interface. And if there is no formation of a dense layer of adsorbed charged molecules of the monomer or its intermediates on the interface, than loose structures form as a result of heterophase process in aqueous solution and they are adsorbed on the interfacial surface. It happens at aniline polymerization in presence of DPA.

Thus, we have shown that introduction of a slight quantity of substituted aromatic amines lets us to regulate the rate of heterophase polymerization of aniline and the quality of the forming polyaniline layers. This also determines the characteristics of composite materials obtained directly from the solutions of PS with dispersed polyaniline microfilms. As the used organic phase, was obtained directly during the heterophase synthesis, electrophysical and morphological characteristics

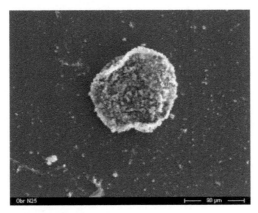

FIGURE 10.4   Electron microscopic image of polyaniline capsules, forming in heterophase conditions after addition of 5% PDA to the reaction mixture.

of polyaniline remain also after the allocation of samples. According to scanning electron microscopy it was found that during the process of obtaining of composite films via casting on a cellophane substrate they retain the layered structure and uniformity of microlayers distribution in volume. Comparison of IR spectra of the surface of composite films, taken at different points from both sides showed that all the spectra are absolutely identical which indicates the homogeneity of the composite structure.

Electrophysical and mechanical characteristics of the obtained composite films were studied. The obtained results are shown in the Table 10.1.

TABLE 10.1   Electrophysical and Mechanical Characteristics of Composite Films

Sample	Content of PA, %	Electroconductivity, Sm/cm	Tensile strength, MPa
Polystyrene (PS)	0	–	58
Polyaniline/PS	6.5	0.02	56
Polyaniline/PS	13	0.15	50
Polyaniline/PS	25	0.6	45
Polyaniline + 5% p-PDA/PS	25	0.9	46
Polyaniline + 2% m-PDA/PS	25	0.03	39

As is seen from the Table 10.1, with the increase of PA content the electrophysical characteristics significantly increase with an insufficient decrease of strength characteristics. The highest conductivity (0.9 Sm/cm) along with maintaining high strength characteristics (46 MPa) belongs to composite films, obtained directly during the oxidation of aniline at the interface in presence of 5 wt.% of p-PDA.

KEYWORDS

- diphenylamine
- heterophase polymerization
- polyaniline
- polymer–polymer compositions

REFERENCES

1. Orlov, A. V., Kiseleva, S. G., Karpacheva, G. P. Peculiarities of polyaniline polymerization in presence of added substrate. Polymer Science. A. 2000, V. 42, № 12, 1292–1297.
2. Karpacheva, G. P., Orlov, A. V., Kiseleva, S. G., Ozkan, S. Zh., Iourchenko, O. Iu. New approaches to the synthesis of electroactive polymers. Electrochemistry. 2004, V. 40, № 3, 346–351.
3. Orlov, A. V., Kiseleva, S. G., Karpacheva, G. P. Interpretation of peculiarities of interfacial aniline polymerization in terms of the model of the double electric layer. Polymer science. A. 2008, V. 50, № 10, 1749–1757.

CHAPTER 11

NOVEL HYBRID METAL-POLYMER NANOCOMPOSITE MAGNETIC MATERIAL BASED ON POLYPHENOXAZINE AND CO-NANOPARTICLES

S. ZH. OZKAN, G. P. KARPACHEVA, and I. S. EREMEEV

A.V. Topchiev Institute of Petrochemical Synthesis, Russian Academy of Sciences, 29, Leninsky prospect, 119991, Moscow, Russia

CONTENTS

ABSTRACT

During the IR heating of polyphenoxazine (PPhOA) in presence of cobalt (II) acetate $Co(CH_3CO_2)_2 \cdot 4H_2O$ in the inert atmosphere at the temperature $T = 500–650°C$ the growth of the polymer chain via condensation of phenoxazine oligomers happens simultaneously with the dehydrogenation with the formation of conjugated C=N bonds. Hydrogen emitted during these processes contributes to the reduction of CO_2^+ to $Co°$. As a result the nanostructured composite material in which Co-nanoparticles are dispersed into the polymer matrix is formed. According to TEM Co-nanoparticles have size 4 < d < 14 nm. The investigation of magnetic and thermal properties of Co/PPhOA nanocomposite has shown that the obtained nanomaterial is superparamagnetic and thermally stable.

11.1 INTRODUCTION

Metal-polymer nanocomposites combine useful properties of polymers and metal nanoparticles. Materials, based on polymers with a system of polyconjugation and magnetic nanoparticles attract a special attention due to their unique physico-chemical properties. Metal-polymer nanomaterials based on polymers with a system of conjugation are promising for application in organic electronics and electrorheology, creating microelectromechanical systems, supercapacitors, sensors, solar cells, displays, etc. Inclusion of magnetic nanoparticles into the nanocomposites makes them promising for use in systems of magnetic recording of information, for creating of electromagnetic screens, contrasting materials for magnetic resonance tomography, etc.

In the current work a method of synthesis of the hybrid metal-polymer nanocomposite based on polyphenoxazine (PPOA) and Co-nanoparticles was developed for the first time. PPOA is a half-ladder heterocyclic polymer, containing both nitrogen and oxygen atoms, which are included into the whole system of polyconjugation. The molecular mass of PPhOA is $M_w = 3.7 \times 10^4$. The selection of the polymer is described by its high thermal stability (up to 400°C in air and in the inert atmosphere at 1000°C the residue is 51%) [1].

The use of IR radiation for chemical and structural modifications instead of the generally used thermal impact is caused by the fact that due to the transition of the system into the vibrationally excited state it becomes possible to increase the rate of chemical reactions and thus to reduce the process time significantly [2–6].

During the IR heating of PPhOA in presence of cobalt (II) acetate $Co(CH_3CO_2)_2 \cdot 4H_2O$ in the inert atmosphere at the temperature T = 500–650°C the growth of the polymer chain via condensation of phenoxazine oligomers happens simultaneously with the dehydrogenation with the formation of conjugated C=N bonds.

According to IR spectroscopy data the growth of the polymer chain is proved by the reduction of intensity of absorption band at 739 cm^{-1},

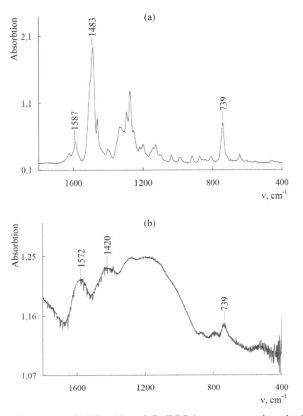

relating to nonplanar deformation vibrations of δ_{C-H} bonds of 1,2-substituted benzene ring of the end groups, that is, the number of polymers end groups significantly reduces. Absorption bands at 869 and 836 cm⁻¹ are caused by nonplanar deformation vibrations of δ_{C-H} bonds of 1,2,4-substituted benzene ring (Figure 11.1). Presence of these bands indicates, that the growth of the polymer chain proceeds via C–C – connection type into para-positions of phenyl rings with respect to nitrogen [1].

FIGURE 11.1 IR spectra of PPOA (a) and Co/PPOA nanocomposite, obtained at 500°C while heating for 10 min, (b) in the absorption range of nonplanar deformation vibrations of δ_{C-H} bonds of aromatic rings.

The formation of C=N bonds is proved by the shift and broadening of the bands at 1587 and 1483 cm^{-1}, corresponding to stretching vibrations of v_{C-C} bonds in aromatic rings. Absorption band at 3380 cm^{-1}, corresponding to the stretching vibrations of v_{N-H} bonds in phenyleneamine structures practically disappears. The absorption band at 3420 cm^{-1}, associated with water appears (Figure 11.2).

Hydrogen emitted during these processes contributes to the reduction of CO_2^+ to $Co°$. As a result the nanostructured composite material in which Co-nanoparticles are dispersed into the polymer matrix is formed. The reflection bands of β-Co-nanoparticles with a cubic face-centered lattice are identified in the diffraction pattern of the nanocomposite in the range of scattering angles $2\theta = 68.35°$, $80.65°$ (Figure 11.3). According to TEM Co-nanoparticles

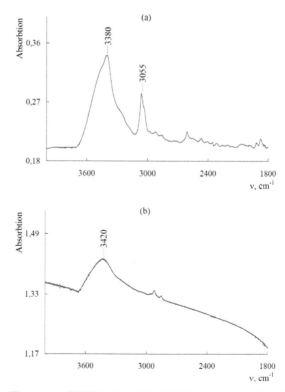

FIGURE 11.2   IR spectra of PPOA (a) and Co/PPOA nanocomposite, obtained at 500°C while heating for 10 min, (b) in the absorption range of stretching vibrations of v_{N-H} and v_{C-H} bonds.

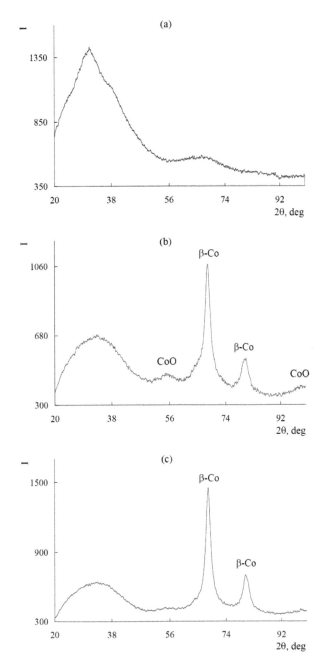

FIGURE 11.3 Diffraction patterns of PPOA (a) and Co/PPOA nanocomposite, obtained at 450 (b) and 500°C while heating for 10 min (c).

has size 4 < d < 14 nm (Figure 11.4). According to AAS the content of cobalt in Co/PPOA nanocomposite, obtained at 550°C is 22.6 wt.%.

Metal-polymer Co/PPhOA nanocomposite is a black powder, insoluble in organic solvents.

It was found that at temperatures below 500°C along with the Co-nanoparticles, the nanoparticles of CoO are also present in the nano-composite in the range of scattering angles $2\theta = 55.61°, 65.32°, 99.05°$ (Figure 11.3b). The increase of heating time in the range of 5–30 min does not lead to the complete reduction of CoO to Co. At low concentrations of Co while loading, only nanoparticles of metallic Co are registered, but when [Co] = 30 wt.% the CoO nanoparticles prevail.

Investigation of magnetic properties at room temperature showed that the obtained Co/PPOA nanocomposites have hysteresis magnetization reversal (Figure 11.5). The values of the main magnetic characteristics of the nano-composite, obtained at 500 and 650°C, are shown in Table 11.1. It was found that after the increase of the temperature the coercive force H_C decreases together with the squareness coefficient of the hysteresis loop k_S, which is the ratio of the residual magnetization M_R to the saturation magnetization M_S. It happens due to the increase of the part of magnetic nanoparticles in the nano-composite through the reduction of CoO to Co at high temperatures. Hybrid Co/PPOA nanocomposite is superparamagnetic. The obtained value $M_R/M_S = 0.116$–0 is characteristic of uniaxial, single-domain particles [5, 6].

Co/PPOA nanocomposite is characterized by the high thermal stability (Figure 11.6). 7% weight loss happens due to the presence of moisture in the nanocomposite, which is proved by DSC data (Figure 11.7). In the

FIGURE 11.4 Microphotograph of Co/PPOA nanocomposite.

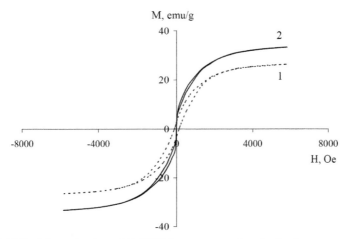

FIGURE 11.5 Magnetization of Co/PPOA nanocomposite, obtained at temperature 500 (1) and 650°C while heating for 10 min (2), as a function of the applied magnetic field at room temperature.

TABLE 11.1 Magnetic Characteristics of Co/PPOA Nanocomposite

Temperature of the sample, °C	H_C, Oe	M_S, emu/g	M_R, emu/g	M_R/M_S
500	134	26.33	3.05	0.116
650	0	33.33	0	0

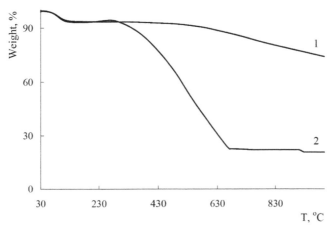

FIGURE 11.6 Weight decrease of Co/PPOA nanocomposite after heating to 1000°C at heating rate 10°C/min in nitrogen flow (1) and in air (2).

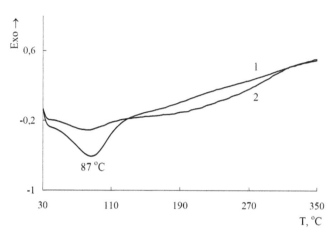

FIGURE 11.7   DSC thermograms of Co/PPOA nanocomposite while heating in nitrogen flow up to 350°C at heating rate 10°C/min (1 – first heating, 2 – second heating).

DSC thermogram of Co/PPOA nanocomposite there is an endothermic peak in this range of temperatures. After reheating this peak is absent.

After the removal of moisture the mass of the Co/PPOA nanocomposite does not change up to 300°C. In the inert atmosphere there is a gradual weight loss and at 1000°C the residue is 75%.

ACKNOWLEDGEMENT

The work has been supported by the Russian Foundation for Basic Research, project № 14-03-31556 mol_a.

KEYWORDS

- **Co-nanoparticles**
- **IR heating**
- **magnetic material**
- **metal-polymer nanocomposite**
- **polyphenoxazine**

REFERENCES

1. Ozkan, S. Zh., Karpacheva, G. P., Bondarenko, G. N. Polymers of phenoxazine: synthesis, structure. Proceedings of the Academy of Sciences. Chemical series. 2011, № 8, 1625–1630.
2. Ozkan, S. Zh., Kozlov, V. V., Karpacheva, G. P. Novel Nanocomposite based on Polydiphenylamine and Nanoparticles Cu and Cu_2O. J. Balkan Tribological Association. 2010, Book 3, V. 16, № 3, 393–398.
3. Ozkan, S. Zh., Karpacheva, G. P. Novel Composite Material Based on Polydiphenylamine and Fe_3O_4 Nanoparticles. Organic Chemistry, Biochemistry, Biotechnology and Renewable Resources. Research and Development. Volume 2 – Tomorrow and Perspectives. Editors: G. E. Zaikov, O. V. Stoyanov, E. L. Pekhtasheva. Nova Science Publishers, Inc. New York. 2013, Chapter 8, 93–96.
4. Ozkan, S. Zh., Dzidziguri, E. L., Krpacheva, G. P., Bondarenko, G. N. Metal-polymer nanocomposites based on polydiphenylamine and copper nanoparticles: synthesis, structure and properties. Russian Nanotechnologies. 2011, Vol. 6, № 11–12, 78–83.
5. Karpacheva, G., Ozkan, S. Polymer-metal hybrid structures based on polydiphenylamine and Co-nanoparticles. Procedia Materials Science. 2013, Vol. 2, 52–59.
6. Ozkan, S. Zh., Dzidziguri, E. L., Chernavskii, P. A., Krpacheva, G. P., Efimov, M. N., Bondarenko, G. N. Metal-polymer nanocomposites based on polydiphenylamine and cobalt nanoparticles. Russian Nanotechnologies. 2013, Vol. 8, № 7–8, 34–40.

CHAPTER 12

USE OF MICROSIZED FERROCOMPOSITES PARTICLES FOR IMMOBILIZATION OF BIOLOGICALLY ACTIVE COMPOUNDS

LUBOV KH. KOMISSAROVA and VLADIMIR S. FEOFANOV

N.M. Emanuel Institute of Biochemical Physics of the Russian Academy of Sciences, Kosygin St. 4 117977 Moscow, Russia, Tel.: 8(495)9361745 (office), 8(906)7544974 (mobile); Fax: (495)1374101; E-mail: komissarova–lkh@mail.ru

CONTENTS

ABSTRACT

We have worked out new methods to modify the surface of the ferrocomposites microsized particles (iron, magnetite, iron–carbon) by albumin, gelatin or dextran and have been studied immobilization hemoglobin and barbiturates (sodium phenobarbital and barbituric acid) and immobilization and dynamics of L-borophenilalanin (L-BPA) desorption. The optimal ferrocomposite types and the methods modifications their surface are suggested as sorbents for extracorporal detoxification of patients blood and purification of donor conserved blood and as carriers for magnetically guided targeted delivery of L-BPA at Boron Neutron Capture of Tumor Therapy (BNCT).

12.1   AIMS AND BACKGROUND

Magnetic nano- and microsized particles can be used for various biomedical applications: cell separation, immobilization of enzymes and viruses, detoxification of biological liquids, magnetic drug targeting und others [1–5]. The most widespread for neutron capture therapy (NCT) have become compounds with ^{10}B (BNCT). The two boron containing compounds, one of them L-borophenilalanin (L-BPA) are used in clinical practice [6]. The aim of the research is to work out new methods to modify the surface of different chemical content microsized ferrocomposites particles by biocompatible materials for the following immobilization of biologically active compounds and to evaluate possibility to use them as sorbents for extracorporal detoxification of patients blood and conserved donor blood purification from free hemoglobin and barbiturates by the method of magnetic separation and as carriers for magnetically guided targeted delivery of L-BPA at BNCT.

12.2   EXPERIMENTAL PART

We have studied composites: iron-silica ($FeSiO_2$) of content: 50%Fe, 50% SiO_2 iron-carbon (FeC) of content 44%Fe, 56%C, iron-carbon-silica ($FeCSiO_2$ of content 50%Fe, 40%C and 10% SiO_2, iron of content: 90% restored iron and 10%, Fe_3O_4 sized 0.02–0.1 mkm, obtained by plasmo-chemical method [5] magnetite (Fe_3O_4) sized 0.1–0.5 mkm, synthesized by chemical method [2]. Diameter of ferrocomposites microparticles in 0,6% of albumin solution: 1–2 mkm² The powders of ferromagnetics were treated (suspension in distillate water) by ultrasonic waves (frequency 22 kHz) in order to eliminate aggregation and to attain a homogenous distribution of the particles in suspension. The particles' surface, besides the same of composites $FeSiO_2$ and $FeCSiO_2$, which is biocompatible, was covered by albumin, or gelatin, or dextran. Carboxylate-magnetic particles were obtained by bovine albumin or gelatin coating with the following aldehydes modification, aldehyde-magnetic particles – dextran coating with $NaJO_4$ activation. We coated the particles by mixing a suspension of particles and albumin or gelatin or dextran with Mm 67,000 Da (Sigma) using ultrasound, with the following 1 h incubation at 20°C, separation of particles on Sm-Co magnet with inductance of 0.1–0.15 Tl. Thereafter the particles were incubated in the modificator solutions: formaldehyde (Russia), or glutaraldehyde, or $NaJO_4$ (Sigma) and washed with distillate water. Surface-modified particles were kept at 10% concentration in physiological solution.

We have used bovine hemoglobin (Sigma), which contained to 75% methemoglobin and to 25% oxyhemoglobin. Immobilization of hemoglobin and barbiturates (Russia) was carried out by 30 sec incubation with the suspension of particles in physiological solution and in a model biological liquid (0.6% albumin in physiological solution) at 20°C (pH 7.4) at different weight ratios of composite/substance: 10, 20, 50 and thereafter the particles were separated on Sm-Co magnet. We have chosen incubation time 30 sec according to the length of contact of biological liquids with suspension of magnetic microparticles in the device for extracorporal detoxification of blood by the method of magnetic separation [1]. Concentrations of compounds in the solutions were measured by differential visual and UV-spectroscopy. The sorption efficiency of ferrocomposites was evaluated as the ratio of the quantity of the adsorbed substance to its initial amount (w/w), expressed in % and in mg/g composite (absorptive capacity) for a certain weight ratio of composite /substance.

Immobilization L-BPA (Lachema) carried out by 10 min incubation with the suspension of particles in acidified water solutions at different weight ratios of composite/L-BPA. The dynamics of L-BPA desorption was studied by incubation of magnetic preparations with immobilized of L-BPA with fresh aliquots of 0.6% albumin at 37°C and by following registration of supernatant absorption UV-spectra. Concentration of desorbed BPA was evaluated on the Calibri curve.

12.3 RESULTS AND DISCUSSION

12.3.1 IMMOBILIZATION OF HEMOGLOBIN AND BARBITURATES

The maximal sorption efficiency to hemoglobin on unmodified ferrocomposites particles showed for Fe_3O_4 and Fe-particles: 40.0 mg/g and 37.8 mg/g, respectively. The results of sorption efficiency of modified ferrocomposites particles to hemoglobin are presented in Table 12.1.

Table 12.1 shows that the sorption efficiency of magnetite to hemoglobin after covering the surface by gelatin is decreasing, glutar-modification leads to its further decrease. Sorption efficiency of magnetite does not exchange practically after formaldehyde modification by gelatin-covered particles. The same character of sorption efficiency of gelatin and albumin-modified particles to hemoglobin is discovered for Fe and FeC particles: sorption efficiency of is decreasing after covering of particles by proteins, the glutar-modification leads to its further decrease and does not change after formaldehyde modification. Immobilization of hemoglobin on aldehyde-modified particles, evidently is due to forming hydrogen connections between carboxylate groups of proteins and amino-groups of hemoglobin. Decrease of hemoglobin adsorption efficiency of particles with glutar-modified surface is predetermined, obviously, by a stereochemical factor.

The results on immobilization of hemoglobin on modified ferrocomposites particles are shown in and Table 12.2 and Figure 12.3.

Figure 12.2 and Table 12.2 demonstrate that iron particles covered by dextran and activated by $NaJO_4$ have shown maximal sorption efficiency to hemoglobin, which is, obviously, accounted for forming hydrogen connections between aldehydes groups of dextran and amino-groups of hemoglobin.

Sorption efficiency of modified ferrocomposites to hemoglobin in a model biological liquid (0.6% albumin in physiological solution) is presented in Table 12.3.

Results on sorption efficiency of modified particles magnetite and iron to hemoglobin in model biological liquid (Table 12.3) showed, that maximal absorptive capacity manifested Fe-particles, modified by albumin (32.6 mg/g) and dextran (25.0 mg/g). These meanings are lower than

TABLE 12.1 Sorption Efficiency of Gelatin-Modified Fe_3O_4-Particles to Hemoglobin at Different Weight Ratios Fe_3O_4/Hb in Physiological Solution at pH 7.4

Types of composites, m	Sorption, average ± SD (%)						
	Absorptive capacity, average ± SD (mg/g)						
	Unmodified		Gelatin-covered		Gelatin + glutarald-modified		Gelatin + formal-modified
Fe_3O_4/Hb	20	50	20	50	20	50	20
	41.6 ± 4.7	70.4 ± 9.3	34.9 ± 4.9	57.8 ± 6.1	26.9 ± 3.6	47.8 ± 5.9	41.0 ± 4.2
	20.8 ± 2.4	14.1 ± 4.7	17.4 ± 2.4	11.6 ± 1.2	13.4 ± 1.8	9.6 ± 1.2	20.5 ± 2.1

TABLE 12.2   Sorption Efficiency of Modified Ferrocomposites to Hemoglobin
(Composite/Hb, w/w=20) in Physiological Solution at pH 7.4

	Sorption, average ± SD (%)			
	Absorptive capacity, average ± SD (mg/g)			
Types of composites	Fe + gelatin + formald.	Fe_3O_4 + gelatin + formald.	Fe + dextran + $NaJO_4$	Fe + albumin + formald.
	38.0 ± 4.6	41.0 ± 4.2	94.5 ± 11.3	68.4 ± 7.5
	19.0 ± 2.3	20.5 ± 2.1	47.2 ± 5.6	34.2 ± 3.8

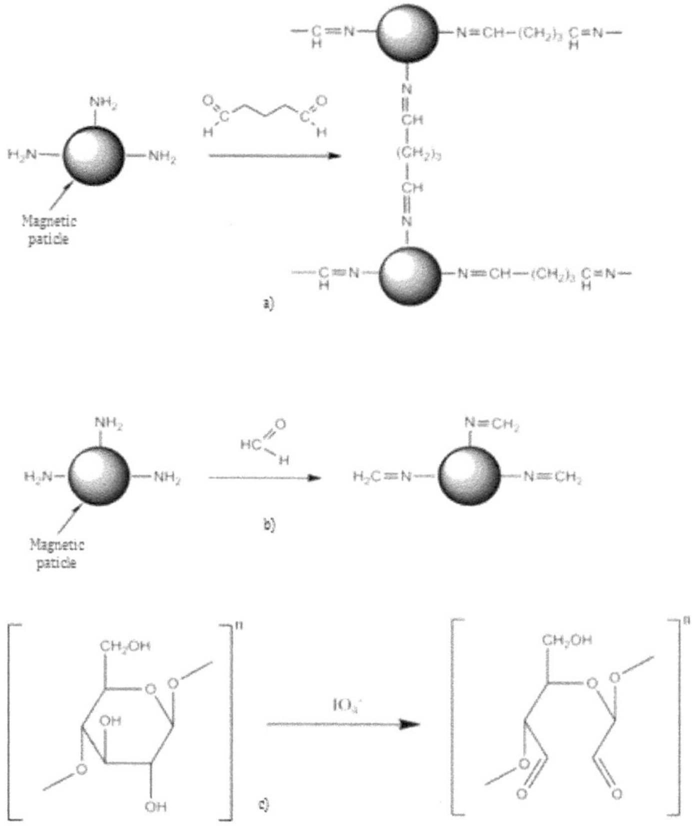

FIGURE 12.1   Modification of proteins (albumin or gelatin) NH_2-groups by glutar (a)
and formaldehydes (b), activation of dextran OH-groups with $NaJO_4$.

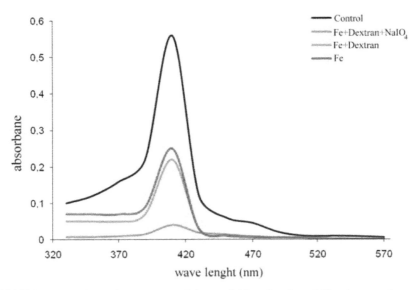

FIGURE 12.2 Absorption spectra of hemoglobin after immobilization on dextran-modified Fe-particles (Fe/Hb, w/w=20) in physiological solution at pH 7.4.

TABLE 12.3 Sorption Efficiency of Modified Ferrocomposites to Hemoglobin (Composite/Hb, w/w=10) in 0.6% Albumin in Physiological Solution at pH 7.4

	Sorption, average ± SD (%)			
	Absorptive capacity, average ± SD (mg/g)			
Types of composites	Fe_3O_4 + gelatin + formald.	Fe + gelatin + formald.	Fe + albumin + formald.	Fe + dextran + $NaJO_4$
	8.2 ± 1.7	16.3 ± 2.4	32.6 ± 3.8	25.0 ± 3.2
	8.2 ± 1.7	16.3 ± 2.4	32.6 ± 3.8	25.0 ± 3.2

those in physiological solution. This can be explained by decreasing of the sorption processes velocity due to increasing of solution viscosity. In fact, the sorption efficiency increases at increasing the incubation time from 30 sec to 60 sec. The interesting results on sorption efficiency of hemoglobin, carboxyhemoglobin and methemoglobin on gelatin-modified Fe-particles have been claimed in donor blood. These meanings are equal: 60.7%, 52.9%, and 22.5%, accordingly.

It is important emphasize that adsorption of albumin on an modified particles reached to 40% for all composites types, after modification of composites surface adsorption of albumin was not more than 10%.

The sorption efficiency results of different chemical content modified ferrocomposites to phenobarbital in physiological solution are represented in Figures 12.3 and 12.4 and in Tables 12.4 and 12.5.

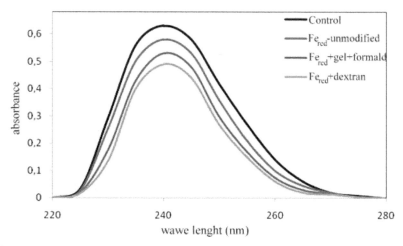

FIGURE 12.3 Absorption spectra of phenobarbital after immobilization on different chemical content Fe-particles (Fe/PhB, w/w=20) in physiological solution at pH 7.4.

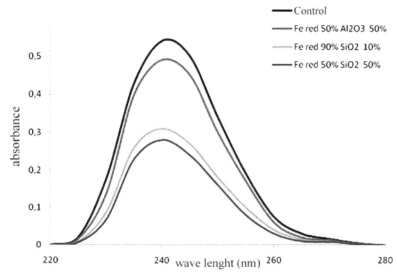

FIGURE 12.4 Absorption spectra of phenobarbital after immobilization on different chemical content Fe-particles (Fe/PhB, w/w=20) in physiological solution at pH 7.4.

TABLE 12.4 Sorption Efficiency of Different Chemical Content Modified Ferrocomposites to Phenobarbital (PhB), (Composite/PhB, w/w=20) in Physiological Solution at pH 7.4

	Sorption, average ± SD (%)			
	Absorptive capacity, average ± SD (mg/g)			
Types of composites	Fe unmodified	Fe + gel + formald.	Fe + albumin + formald.	Fe-Al$_2$O$_3$
	18.9 ± 1.8	36.2 ± 2.3	51.4 ± 5.8	14.2 ± 1.5
	9.4 ± 0.9	18.1 ± 1.2	25.7 ± 2.9	7.0 ± 0.7

TABLE 12.5 Sorption Efficiency of Different Chemical Content Ferrocomposites to Phenobarbital (PhB), (Composite/PhB, w/w=20) in Physiological Solution at pH 7.4

	Sorption, average ± SD (%)			
	Absorptive capacity, average ± SD (mg/g)			
Types of composites	Fe unmodified	Fe + Dex	Fe-Silica (Fe 90%, SiO$_2$ 10%)	Fe-Silica (Fe 50%, SiO$_2$ 50%)
	18.9 ± 1.8	15.9 ± 1.4	42.6 ± 4.2	48.1 ± 6.4
	9.4 ± 0.9	7.9 ± 0.7	21.3 ± 2.1	24.1 ± 3.2

Table 12.4 and shows that modification of Fe-microparticles surface by albumin led to considerable increase of phenobarbital immobilization: from 18.9% to 51.4%. The immobilization is realized, probably, by means of conjugation of phenobarbital with carboxylate- groups of albumin. In Figures 12.3 and 12.4 are shown absorption spectra of phenobarbital after immobilization on different chemical content microparticles.

Maximal meanings of sorption efficiency of phenobarbital have demonstrated Fe-silica composites. Formation of hydrogen connections plays, apparently, a prevailing role in immobilization phenobarbital on Fe-silica composites (Figure 12.4 and Table 12.5).

In Figure 12.5 are shown spectra of barbituric acid after immobilization on FeC SiO$_2$ microparticles at different weight ratios composite/BA. The maximal sorption efficiency of barbituric acid was found for FeC SiO$_2$ composite of content: 50% Fe, 40%C, 10%SiO$_2$. The meanings of sorption and absorptive capacity of barbituric acid for this composite:

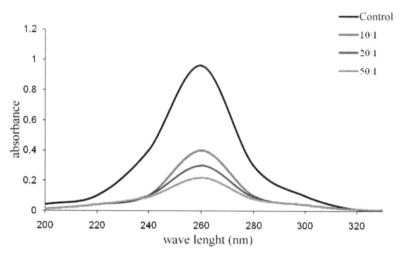

FIGURE 12.5 Absorption spectra of barbituric acid after immobilization on FeC SiO₂ particles at weight ratios composite/ BA= 10, 20, 50.

75.0% and 58.0 mg/g at weight ratio $FeCSiO_2/BA$ 50 and 10, accordingly. Apparently, it occurs physical adsorption of barbituric acid in microporous of composite.

12.3.2 IMMOBILIZATION AND DESORPTION OF L-BOROPHENILALANIN

In Figure 12.6 are shown spectra of L-BPA after immobilization on FeC microparticles at different weight ratios composite/L-BPA. Apparently immobilization occurs by physical adsorption into porous of composite. The highest absorption capacity of L-BPA for this composite 78.0 mg/g was detected at weight ratio composite/L-BPA equal 5. The maximal adsorption capacity of L-BPA 160.0 mg/g was reached for dextran-modified iron-particles.

The desorption of L-BPA at λ225 nm from FeC composite and dextran-modified Fe-particles are presented in Figures 12.7 and 12.8. Analysis of the results on desorption has shown that quantity of the desorbed L-BPA from magnetic operated preparations is enough to create therapeutic

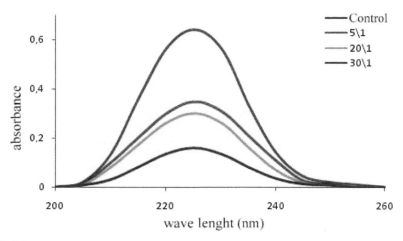

FIGURE 12.6 Absorption spectra of L-BPA after immobilization on FeC microparticles at different weight ratios composite/L-BPA.

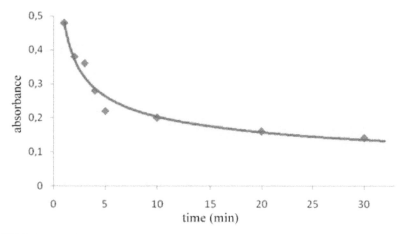

FIGURE 12.7 The dynamics of L-BPA desorption (λ225 nm) from FeC composite in 0.6% albumin (T 37°C, pH 7.4).

concentration of boron atoms in tumor. Nevertheless it is required to continue investigations in order to chose the optimum ferrocomposites types with more longer of desorption time for working out magnetic operated preparations of L-BPA on their basis.

Spectrophotometric study of the reaction interaction L-BPA with dextran in the water solutions showed its conjugation with dextran (Figure 12.9).

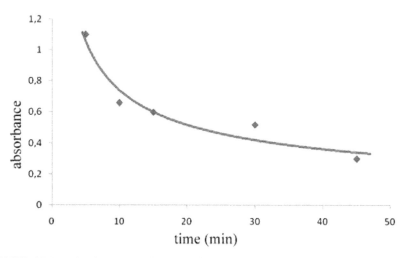

FIGURE 12.8 The dynamics of L-BPA desorption (λ225 nm) from dextran-modified Fe-particles in 0.6% albumin (T 37°C, pH 7.4).

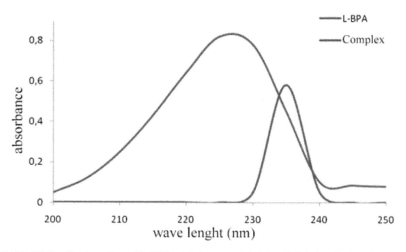

FIGURE 12.9 Conjugation of L-BPA with dextran: T 20°C, pH 7.0, incubation time 1 min.

12.4 SUMMARY

We have worked out new methods to modify the surface of the ferrocomposites microsized particles (iron, magnetite, iron-carbon) by albumin, gelatin or dextran and have been studied their sorption efficiency and the same for composite iron-silica and iron-carbon-silica to bovine hemoglobin and barbiturates: sodium phenobarbital and barbituric acid. Optimal Fe-composites for hemoglobin immobilization are by albumin and dextran modified microsized Fe-particles; for phenobarbital: albumin-modified Fe-particles and Fe-silica composite; for barbituric acid:FeC-silica composite. These ferrocomposites can be recommended for use as sorbents for extracorporal detoxification of patients' blood and purification of conserved donor blood from free hemoglobin and barbiturates by the method of magnetic separation. Dextran modified microsized Fe-particles are perspective as carriers for magnetically guided targeted delivery of L-BPA at Boron Neutron Capture of Tumor Therapy.

KEYWORDS

- barbiturates
- hemoglobin
- immobilization of biologically active compounds
- L-borophenilalanin
- microsized particles of ferrocomposites
- modified surface
- sorption efficiency

REFERENCES

1. Komissarova, L. Kh., Filippov, V. I., Kuznetsov, A. A. In the Proceeding of the 1st Symposium on Application of biomagnetic carriers in medicine (in Rus.), Moscow, Science (Nauka) Publishing House, 2002, 68–76.
2. Brusentsov, N. A., Bayburtskiy, F. S., Komissarova, L. Kh. et al. Biocatalytic Technology and Nanotechnology. Nova science Publishers, New York, 2004, 59–66.
3. Yanovsky, Y. G., Komissarova, L. Kh., Danilin, A. N. et al. Solid State Phenomena. 2009, 152–153, 403–406.
4. Zhiwei Li, Chao Wang, Liang Cheng et al. Biomaterials. 2013, No. 4, 9160–9170.
5. Kutushov, M. W., Komissarova, L. Kh., Gluchoedov, N. P. Russian patent. No 210952, 1998.
6. Bunis, R. J., Riley, K. J., Marling, O. K. In: Research and Development in Neutron Capture Therapy. Ed. Monduzzi. Bologna, 2002, 405.

CHAPTER 13

THE AGGREGATION PROCESS AS LARGE-SCALE DISORDER CAUSE IN NANOCOMPOSITES POLYMER/ ORGANOCLAY

G. V. KOZLOV,[1] G. E. ZAIKOV,[2] and A. K. MIKITAEV[1]

[1]Kh.M. Berbekov Kabardino-Balkarian State University, Chernyshevsky St., 173, 360004 Nalchik, Russian Federation; E-mail: i_dolbin@mail.ru

[2]N.M. Emanuel Institute of Biochemical Physics of Russian Academy of Sciences, Kosygin St., 4, 119991 Moscow, Russian Federation; E-mail: chembio@sky.chph.ras.ru

CONTENTS

ABSTRACT

It has been shown that organoclay platelets aggregation in "packets" (tactoids) results in large-scale disorder intensification that reduces nanofiller anisotropy degree. In its turn, this factor decreases essentially nanocomposites reinforcement degree. The interfacial adhesion role in anisotropy level definition has been shown.

13.1 INTRODUCTION

Organoclay belongs to anisotropic nanofillers, for which their anisotropy degree, that is, ratio α of organoclay platelets (aggregates of platelets) length to thickness, has large significance [1]. The reinforcement degree E_n/E_m of nanocomposites polymer/organoclay can be estimated according to the equation [1]:

$$\frac{E_n}{E_m} = 1 + 2\alpha C_a \varphi_n \qquad (1)$$

where E_n and E_m are elasticity moduli of nanocomposite and matrix polymer, respectively, C_a is an orientation factor, which is equal for organoclay to about 0.5 [1], φ_n is organoclay volume content.

In case of organoclay platelets aggregation, that is, their "packets" (tactoids) formation [2] such "packets" thickness increasing takes place in comparison with a separate platelet, that results in length/thickness ratio α reduction at platelet constant length and, as consequence, nanocomposites reinforcement degree decreasing is realized according to the Eq. (1). The present work purpose is the analytical study of organoclay aggregation, that is, large-scale disorder, influence on reinforcement degree on the example of nanocomposites plasticat of poly(vinyl chloride) – organomodified montmorillonite.

13.2 EXPERIMENTAL PART

The plasticat of poly(vinyl chloride) (PVC) of mark U30–13A, prescription 8/2 GOST 5960–72 was used as a matrix polymer. The modification product of montmorillonite (MMT) of deposit Gerpegezh (KBR, Russian Federation), modified by urea with content of 10 mass% with cation-changing capacity of 95 mg-eq/100 g of clay was applied as a nanofiller. Organoclay contents was varied within the limits of 1–10 mass%.

The nanocomposites PVC-MMT preparation was performed as follows. The components were mixed in a two-speed blender R 600/HC 2500 of firm Diosna, the design of which ensures intensive intermixing in turbulent regime with blends high homogenization and blowing by hot air. After components intensive intermixing the composition was cooled up to temperature 313 K and processed on a twin screw extruder Thermo Haake, model Reomex RTW 25/42, production of German Federal Republic, at temperature 418–438 K and screw rotation speed of 48 rpm.

Sheet nanocomposite was obtained by a hot rolling method at temperature (433±10) K during 5–15 min. The samples in the shape of a two-sided spade with sizes according to GOST 112 62–80 were cut out by punch. Uniaxial tension mechanical tests have been performed on the universal testing apparatus Gotech Testing Machine CT-TCS 2000, production of German Federal Republic, at temperature 293 K and strain rate of $\sim 2 \times 10^{-3}$ s^{-1}.

13.3   RESULTS AND DISCUSSION

The nanofiller initial particles aggregation is the main process, enhancing large-scale disorder level in polymer nanocomposites. For each nanofiller type this process has its specific character, but in case of anisotropic nanofillers (organoclay, carbon nanotubes) application this process always decreases their anisotropy degree, that is, aspect ratio α, that reduces nanocomposites reinforcement degree according to the Eq. (1). Let us consider the physical bases of the value α reduction at organoclay content increasing in the considered nanocomposites. As it is known [3], organoclay platelets number N_{pl} in "packet" (tactoid) can be determined as follows:

$$N_{pl} = 24 - 5.7 b_{\alpha} \qquad (2)$$

where b_a is the dimensionless parameter, characterizing the level of interfacial adhesion polymeric matrix-nanofiller, which is determined with the aid of the following percolation relationship [2]:

$$\frac{E_n}{E_m} = 1 + 11\left(c\varphi_n b_a\right)^{1.7}$$

(3)

where c is constant coefficient, which is equal to 1.955 for intercalated organoclay and 2.90 – for exfoliated one.

In its turn, the value φ_n can be determined according to the well-known formula [2]:

$$\varphi_n = \frac{W_n}{\rho_n}$$

(4)

where W_n is nanofiller mass content; ρ_n is its density, which for nanoparticles is determined as follows [2]:

$$\rho_n = 188\left(D_p\right)^{1/3}, \text{kg/m}^3$$

(5)

where D_p is the initial nanoparticle diameter, which is given in nm.

In case of organoclay parameter D_p is determined as mean arithmetical of its three basic sizes: length, width and thickness, which are equal to 100, 35 and 0.65 nm, respectively [2].

An alternative method of the value N_{pl} estimation gives the following equation [2]:

$$\chi = \frac{N_{pl}d_{pl}}{\left(N_{pl} - 1\right)d_{001} + d_{pl}}$$

(6)

where χ is relative volume content of montmorillonite in tactoid (*effective particle* [4]); d_{pl} is thickness of organoclay separate platelet; d_{001} is interlayer spacing, that is, the distance between organoclay platelets in tactoid, which can be estimated according to the following formula [3]:

$$d_{001} = 1.27b_a, \text{nm}$$

(7)

In its turn, parameter χ is determined as follows [3]:

$$\chi = \frac{\varphi_n}{\varphi_n + \varphi_{if}} \tag{8}$$

where φ_{if} is a relative fraction of interfacial regions in nanocomposite, estimated with the aid of the following percolation relationship [2]:

$$\frac{E_n}{E_m} = 1 + 11\left(\varphi_n + \varphi_{if}\right)^{1.7} \tag{9}$$

The comparison of N_{pl} value calculations according to the Eqs. (2) and (6) showed their close correspondence.

In Figure 13.1 the dependence $N_{pl}(\varphi_n)$ for nanocomposites PVC/MMT is adduced. As one can see, at quite enough small values $\varphi_n \leq 0.05$ fast growth of N_{pl} occurs, that is, strong aggregation of organoclay initial platelets, and at $\varphi_n > 0.05$ the value N_{pl} achieves the asymptotic branch: $N_{pl} \approx 22$. As it was noted above, the reduction of nanofiller anisotropy degree, characterized by parameter α, was defined by its aggregation, the level of which could be characterized by parameter N_{pl}. In Figure 13.2 the dependence $\alpha(N_{pl}^2)$ for the considered nanocomposites, where quadratic shape of dependence was chosen with the purpose of its linearization. As was to be expected, the organoclay anisotropy degree, characterized by the parameter α, reduction is observed at its platelets aggregation,

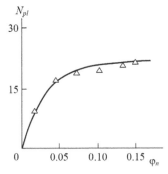

FIGURE 13.1 The dependence of organoclay platelets number per one tactoid N_{pl} on nanofiller volume content φ_n for nanocomposites PVC-MMT.

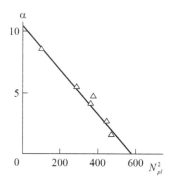

FIGURE 13.2 The dependence of organoclay anisotropy degree α on its platelets number per one tactoid Npl for nanocomposites PVC-MMT.

characterized by the value N_{pl}, enhancement, which is expressed analytically by the following equation:

$$\alpha = 10.5 - 0.018 N_{pl}^2 \tag{10}$$

where the value α was estimated according to the Eq. (1).

Theoretical method of parameter α (α^T) estimation can be obtained as follows. Organoclay aggregates (tactoids) anisotropy degree can be determined according to the equation:

$$\alpha^T = \frac{L_{pl}}{t_{org}} \tag{11}$$

where L_{pl} is organoclay platelet length, which is equal to ~ 100 nm [2], t_{org} is its tactoid thickness.

In its turn, the value t_{org} is determined as follows:

$$t_{org} = d_{001} N_{pl} + 1 \tag{12}$$

Besides, it should be borne in mind, that experimental value α in the Eq. (1) is determined on the basis of reinforcement degree E_n/E_m, that is, on the basis of mechanical tests results. This means, that value α depends on conditions of stress transfer on interfacial boundary polymeric matrix-organoclay, that is, on the parameter b_α value. Then parameter α^T can be determined finally as follows:

$$\alpha^T = \frac{L_{pl}b_\alpha}{1.27b_\alpha N_{pl} + 1} \qquad (13)$$

In Figure 13.3 the comparison of experimental α and calculated according to the Eq. (13) α^T values of organoclay tactoids aspect ratio, characterizing large-scale disorder level for the considered nanocomposites, is adduced. As one can see, good enough correspondence of experiment and theory is obtained (average discrepancy of α and α^T makes up $\sim 9\%$).

The Eqs. (10) and (13) allow to predict reinforcement degree E_n/E_m on organoclay known structural characteristics. In Figure 13.4 the comparison of theoretical curves $E_n/E_m(\varphi_n)$, calculated according to the Eq. (1), where parameter α^T was determined according to the Eqs. (10) and (13), and corresponding the experimental data is adduced. As one can see, a good both qualitative (the theoretical curves are reflected experimental dependence maximum without existence of maximums for parameters N_{pl} and d_{001}) and quantitative correspondence of theory and experiment (their average discrepancy makes up less 2.5 %).

Let consider in conclusion the influence of organoclay platelets aggregation or large-scale disorder on the considered nanocomposites reinforcement degree. In Figure 13.5 the experimental and calculated according to

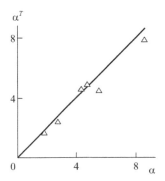

FIGURE 13.3   The comparison of experimental α and calculated according to the Eq. (13) α^T organoclay anisotropy degree for nanocomposites PVC-MMT.

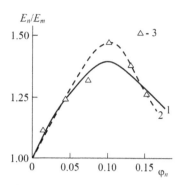

FIGURE 13.4 The comparison of calculated according to the Eq. (1) with usage of the Eq. (10)(1) and Eq. (13)(2) for parameter α^T determination and experimental (3) dependences of reinforcement degree E_n/E_m on nanofiller volume content φ_n for nanocomposites PVC-MMT.

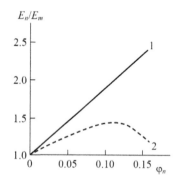

FIGURE 13.5 The comparison of calculated according to the Eq. (1) at the condition of organoclay minimum aggregation (1) and experimental (2) dependences of reinforcement degree E_n/E_m on nanofiller volume content φ_n for nanocomposites PVC-MMT.

the Eq. (1) at organoclay aggregation minimum level (N_{pl}=9.70, b_α=2.51, α=8.6), corresponding to organoclay content W_n=1 mass%, dependences $E_n/E_m(\varphi_n)$ for nanocomposites PVC/MMT are adduced. As it follows from this comparison, both lower values of reinforcement degree and its decay at $W_n > 7$ mass% are due precisely by organoclay platelets aggregation in "packets" (tactoids).

13.4 CONCLUSIONS

Thus, the present work results have demonstrated that organoclay plate-lets aggregation in "packets" (tactoids) results in large-scale disorder enhancement that reduces nanofiller anisotropy degree. In its turn, this factor decreases essentially reinforcement degree (or elasticity modulus) of nanocomposites polymer-organoclay, moreover at large enough organoclay contents (>7 mass%) reinforcement degree reduction at nanofiller content growth is observed. The important role of interfacial adhesion in anisotropy level determination has been shown.

KEYWORDS

- **aggregation**
- **anisotropy**
- **nanocomposite**
- **organoclay**
- **reinforcement degree**
- **tactoid**

REFERENCES

1. Schaefer, D. W., Justice, R. S. How nano are nanocomposites? *Macromolecules*, 2007, vol. 40, no. 24, 8501–8517.
2. Mikitaev, A. K., Kozlov, G. V., Zaikov, G.E. *Polymer Nanocomposites: Variety of Structural Forms and Applications,* New York: Nova Science Publishers, Inc., 2008.
3. Kozlov, G. V., Mikitaev, A. K. *Structure and Properties of Nanocomposites Polymer/Organoclay,* Saarbrücken: LAP LAMBERT Academic Publishing GmbH and Comp., 2013.
4. Sheng, N., Boyce, M. C., Parks, D. M., Rutledge, G. C., Abes, J. I., Cohen, R. E. Multiscale micromechanical modeling of polymer/clay nanocomposites and the effective clay particle, *Polymer*, 2004, vol. 45, no. 2, 487–506.

CHAPTER 14

NANOPOROUS POLYMER/ CARBON NANOTUBE MEMBRANE FILTRATION: THE "HOW-TO" GUIDE TO COMPUTATIONAL METHODS

AREZO AFZALI, SHIMA MAGHSOODLOU, and BABAK NOROOZI

University of Guilan, Rasht, Iran

CONTENTS

ABSTRACT

Membrane filtration is an important technology for ensuring the purity, safety and/or efficiency of the treatment of water or effluents. In this chapter, various types of membranes are reviewed, first. After that, the states of the computational methods are applied to membranes processes. Many studies have focused on the best ways of using a particular membrane process. But, the design of new membrane systems requires a considerable amount of process development including robust methods. Monte Carlo and molecular dynamics methods can specially provide a lot of interesting information for the development of polymer/carbon nanotube membrane processes.

14.1   MEMBRANES FILTRATION

Membrane filtration is a mechanical filtration technique, which uses an absolute barrier to the passage of particulate material as any technology currently available in water treatment. The term "membrane" covers a wide range of processes, including those used for gas/gas, gas/liquid, liquid/liquid, gas/solid, and liquid/solid separations. Membrane production is a large-scale operation. There are two basic types of filters: depth filters and membrane filters.

Depth filters have a significant physical depth and the particles to be maintained are captured throughout the depth of the filter. Depth filters often have a flexuous three-dimensional structure, with multiple channels and heavy branching so that there is a large pathway through which the liquid must flow and by which the filter can retain particles. Depth filters have the advantages of low cost, high throughput, large particle retention capacity, and the ability to retain a variety of particle sizes. However, they can endure from entrainment of the filter medium, uncertainty regarding effective pore size, some ambiguity regarding the overall integrity of the filter, and the risk of particles being mobilized when the pressure differential across the filter is large.

The second type of filter is the membrane filter, in which depth is not considered momentous. The membrane filter uses a relatively thin material with a well-defined maximum pore size and the particle retaining effect takes place almost entirely at the surface. Membranes offer the advantage of having well-defined effective pore sizes, can be integrity tested more easily than depth filters, and can achieve more filtration of much smaller particles. They tend to be more expensive than depth filters and usually cannot achieve the throughput of a depth filter. Filtration technology has developed a well-defined terminology that has been well addressed by commercial suppliers.

The term membrane has been defined in a number of ways. The most appealing definitions to us are the following:

"A selective separation barrier for one or several components in solution or suspension [19]." "A thin layer of material that is capable of separating materials as a function of their physical and chemical properties when a driving force is applied across the membrane."

Membranes are important materials, which form part of our daily lives. Their long history and use in biological systems has been extensively studied throughout the scientific field. Membranes have proven themselves as promising separation candidates due to advantages offered by their high stability, efficiency, low energy requirement and ease of operation. Membranes with good thermal and mechanical stability combined with good solvent resistance are important for industrial processes [1].

The concept of membrane processes is relatively simple but nevertheless often unknown. Membranes might be described as conventional filters but with much finer mesh or much smaller pores to enable the separation of tiny particles, even molecules. In general, one can divide membranes into two groups: porous and nonporous. The former group is similar to classical filtration with pressure as the driving force; the separation of a mixture is achieved by the rejection of at least one component by the membrane and passing of the other components through the membrane (see Figure 14.1). However, it is important to note that nonporous membranes do not operate on a size exclusion mechanism.

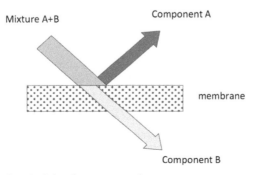

FIGURE 14.1   Basic principle of porous membrane processes.

Membrane separation processes can be used for a wide range of applications and can often offer significant advantages over conventional separation such as distillation and adsorption since the separation is based on a physical mechanism. Compared to conventional processes, therefore, no chemical, biological, or thermal change of the component is involved for most membrane processes. Hence, membrane separation is particularly attractive to the processing of food, beverage, and bioproducts where the processed products can be sensitive to temperature (vs. distillation) and solvents (vs. extraction).

Synthetic membranes show a large variety in their structural forms. The material used in their production determines their function and their driving forces. Typically the driving force is pressure across the membrane barrier (see Table 14.1) [2–4]]. Formation of a pressure gradient across the membrane allows separation in a bolter-like manner. Some other forms of separation that exist include charge effects and solution diffusion. In this separation, the smaller particles are allowed to pass through as permeates whereas the larger molecules (macromolecules) are retained. The retention or permeation of these species is ordained by the pore architecture as well as pore sizes of the membrane employed. Therefore based on the pore sizes, these pressure driven membranes can be divided into reverse osmosis (RO), nanofiltration (NF), ultrafiltration (UF), and microfiltration (MF), are already applied on an industrial scale to food and bioproduct processing [5–7].

TABLE 14.1   Driving Forces and Their Membrane Processes

Driving force	Membrane process
Pressure difference	Microfiltration, Ultrafiltration, Nanofiltration, Reverse osmosis
Chemical potential difference	Pervaporation, Pertraction, Dialysis, Gas separation, Vapor permeation, Liquid Membranes
Electrical potential difference	Electrodialysis, Membrane electrophoresis, Membrane electrolysis
Temperature difference	Membrane distillation

14.1.1 MICROFILTRATION (MF) MEMBRANES

MF membranes have the largest pore sizes and thus use less pressure. They involve removing chemical and biological species with diameters ranging between 100 to 10,000 nm and components smaller than this, pass through as permeates. MF is primarily used to separate particles and bacteria from other smaller solutes [4].

14.1.2 ULTRAFILTRATION (UF) MEMBRANES

UF membranes operate within the parameters of the micro- and NF membranes. Therefore UF membranes have smaller pores as compared to MF membranes. They involve retaining macromolecules and colloids from solution, which range between 2–100 nm and operating pressures between 1 and 10 bar (e.g., large organic molecules and proteins). UF is used to separate colloids, such as, proteins from small molecules such as sugars and salts [4].

14.1.3 NANOFILTRATION (NF) MEMBRANES

NF membranes are distinguished by their pore sizes of between 0.5–2 nm and operating pressures between 5 and 40 bar. They are mainly used for the removal of small organic molecules and di- and multivalent ions. Additionally, NF membranes have surface charges that make them suitable for retaining ionic pollutants from solution. NF is used to achieve separation between sugars, other organic molecules, and multivalent salts on the one hand from monovalent salts and water on the other. Nanofiltration, however, does not remove dissolved compounds [4].

14.1.4 REVERSE OSMOSIS (RO) MEMBRANES

RO membranes are dense semi-permeable membranes mainly used for desalination of seawater [38]. Contrary to MF and UF membranes, RO membranes have no distinct pores. As a result, high pressures are applied to increase the permeability of the membranes [4]. The properties of the various types of membranes are summarized in Table 14.2.

The NF membrane is a type of pressure-driven membrane with properties in between RO and UF membranes. NF offers several advantages such as low operation pressure, high flux, high retention of multivalent anion salts and an organic molecular above 300, relatively low investment and low operation and maintenance costs. Because of these advantages, the applications of NF worldwide have increased [8]. In recent times, research in the application of nanofiltration techniques has been extended from separation of aqueous solutions to separation of organic solvents to homogeneous catalysis, separation of ionic liquids, food processing, etc. [9].

Figure 14.2 presents a classification on the applicability of different membrane separation processes based on particle or molecular sizes. RO process is often used for desalination and pure water production, but it is the UF and MF that are widely used in food and bioprocessing.

While MF membranes target on the microorganism removal, and hence are given the absolute rating, namely, the diameter of the largest pore on the membrane surface, UF/NF membranes are characterized by the nominal rating due to their early applications of purifying biological solutions. The nominal rating is defined as the molecular weight cut-off (MWCO) that is the smallest molecular weight of species, of which the membrane has more than 90% rejection (see later for definitions). The separation mechanism in MF/UF/NF is mainly the size exclusion, which is indicated in the nominal ratings of the membranes. The other separation mechanism

TABLE 14.2   Summary of Properties of Pressure Driven Membranes [4]

	MF	UF	NF	RO
Permeability(L/h. m².bar)	1000	10–1000	1.5–30	0.05–1.5
Pressure (bar)	0.1–2	0.1–5	3–20	5–1120
Pore size (nm)	100–10,000	2–100	0.5–2	< 0.5
Separation Mechanism	Sieving	Sieving	Sieving, charge effects	Solution diffusion
Applications	Removal of bacteria	Removal of bacteria, fungi, virses	Removal of multivalent ions	Desalination

Membrane process
Reverse osomosis
Nano-filtration
Ultrafiltration
Microfiltration
Cloth, Fiber filter
Screen
0.001
0.01
0.1
1.0
10
100
1000
Ionic range
Macromolecular range
Micron particle
Fine particle
Coarse particle

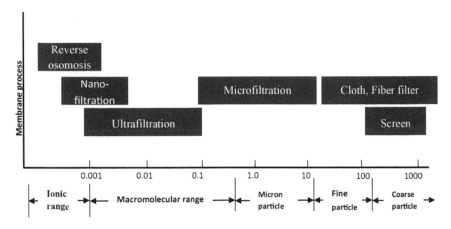

FIGURE 14.2 The applicability ranges of different separation processes based on sizes.

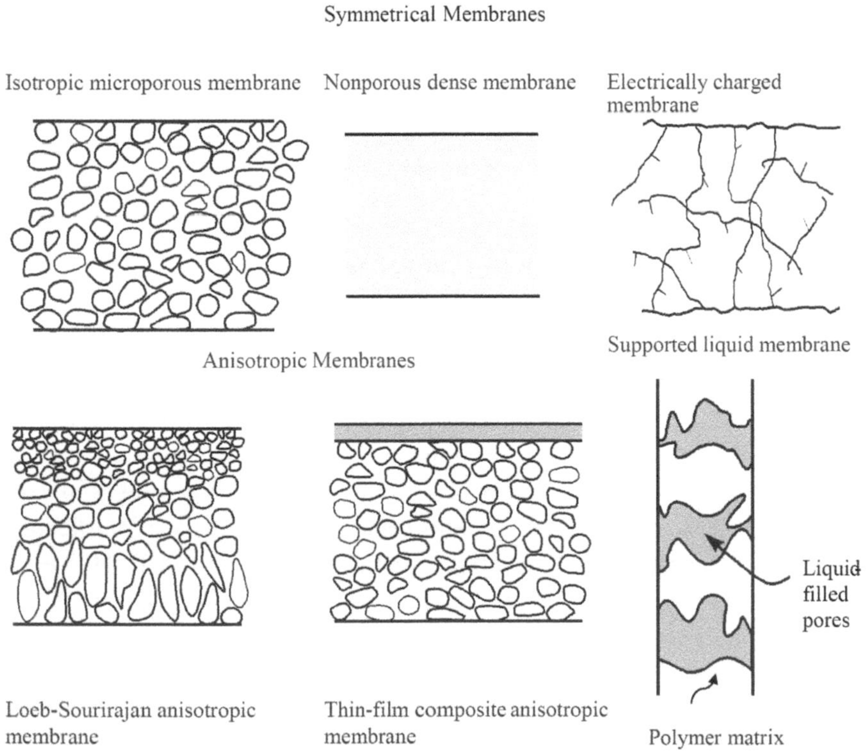

Symmetrical Membranes
Isotropic microporous membrane
Nonporous dense membrane
Electrically charged membrane
Supported liquid membrane
Anisotropic Membranes
Liquid filled pores
Loeb-Sourirajan anisotropic membrane
Thin-film composite anisotropic membrane
Polymer matrix

FIGURE 14.3 Schematic diagrams of the principal types of membranes.

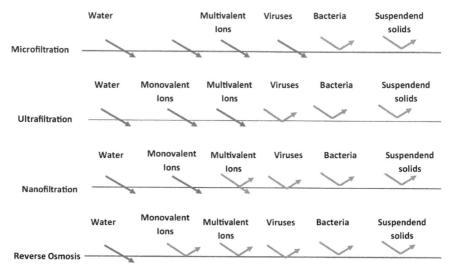

FIGURE 14.4 Membrane process characteristics.

includes the electrostatic interactions between solutes and membranes, which depends on the surface and physiochemical properties of solutes and membranes [5]. Also, The principal types of membrane are shown schematically in Figure 14.4 and are described briefly below.

14.2 THE RELATIONSHIP BETWEEN NANOTECHNOLOGY AND FILTRATION

Nowadays, nanomaterials have become the most interested topic of materials research and development due to their unique structural properties (unique chemical, biological, and physical properties as compared to larger particles of the same material) that cover their efficient uses in various fields, such as ion exchange and separation, catalysis, biomolecular isolation and purification as well as in chemical sensing [10]. However, the understanding of the potential risks (health and environmental effects) posed by nanomaterials hasn't increased as rapidly as research has regarding possible applications.

   One of the ways to enhance their functional properties is to increase their specific surface area by the creation of a large number of nanostructured elements or by the synthesis of a highly porous material.

Classically, porous matter is seen as material containing three-dimensional voids, representing translational repetition, while no regularity is necessary for a material to be termed "porous." In general, the pores can be classified into two types: open pores, which connect to the surface of the material, and closed pores, which are isolated from the outside. If the material exhibits mainly open pores, which can be easily transpired, then one can consider its use in functional applications such as adsorption, catalysis and sensing. In turn, the closed pores can be used in sonic and thermal insulation, or lightweight structural applications. The use of porous materials offers also new opportunities in such areas as coverage chemistry, guest–host synthesis and molecular manipulations and reactions for manufacture of nanoparticles, nanowires and other quantum nanostructures. The International Union of Pure and Applied Chemistry (IUPAC) defines porosity scales as follows:

- Microporous materials 0–2-nm pores
- Mesoporous materials 2–50-nm pores
- Macroporous materials >50-nm pores

This definition, it should be noted, is somewhat in conflict with the definition of nanoscale objects, which typically have large relative porosities (>0.4), and pore diameters between 1 and 100 nm. In order to classify porous materials according to the size of their pores the sorption analysis is one of the tools often used. This tool is based on the fact that pores of different sizes lead to totally different characteristics in sorption isotherms. The correlation between the vapor pressure and the pore size can be written as the Kelvin equation:

$$r_p\left(\frac{p}{p_0}\right) = \frac{2\gamma V_L}{RT\ln\left(\frac{p}{p_0}\right)} + t\left(\frac{p}{p_0}\right) \tag{1}$$

Therefore, the isotherms of microporous materials show a steep increase at very low pressures (relative pressures near zero) and reach a plateau quickly. Mesoporous materials are characterized by a so-called capillary doping step and a hysteresis (a discrepancy between adsorption and desorption). Macroporous materials show a single or multiple adsorption steps near the pressure of the standard bulk condensed state (relative pressure approaches one) [10].

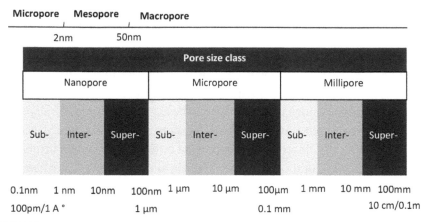

FIGURE 14.5 New pore size classification as compared with the current IUPAC nomenclature.

Nanoporous materials exuberate in nature, both in biological systems and in natural minerals. Some nanoporous materials have been used industrially for a long time. Recent progress in characterization and manipulation on the nanoscale has led to noticeable progression in understanding and making a variety of nanoporous materials: from the merely opportunistic to directed design. This is most strikingly the case in the creation of a wide variety of membranes where control over pore size is increasing dramatically, often to atomic levels of perfection, as is the ability to modify physical and chemical characteristics of the materials that make up the pores [11].

The available range of membrane materials includes polymeric, carbon, silica, zeolite and other ceramics, as well as composites. Each type of membrane can have a different porous structure, as illustrated in Figure 14.6. Membranes can be thought of as having a fixed (immovable) network of pores in which the molecule travels, with the exception of most polymeric membranes [12, 13]. Polymeric membranes are composed of an amorphous mix of polymer chains whose interactions involve mostly Van der Waals forces. However, some polymers manifest a behavior that is consistent with the idea of existence of opened pores within their matrix. This is especially true for high free volume, high permeability polymers, as has been proved by computer modeling, low activation energy of diffusion, negative activation energy of permeation, solubility controlled permeation

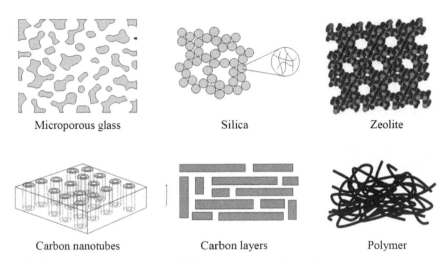

Microporous glass	Silica	Zeolite
Carbon nanotubes	Carbon layers	Polymer

FIGURE 14.6 Porous structure within various types of membranes.

[14, 15]. Although polymeric membranes have often been viewed as non-porous, in the modeling framework discussed here it is convenient to consider them nonetheless as porous. Glassy polymers have pores that can be considered as 'frozen' over short times scales, while rubbery polymers have dynamic fluctuating pores (or more correctly free volume elements) that move, shrink, expand and disappear [16].

Three nanotechnologies that are often used in the filtering processes and show great potential for applications in remediation are:

1. NF (and its sibling technologies: RO, UF, and MF), is a fully developed, commercially available membrane technology with a large number of vendors. NF relies on the ability of membranes to discriminate between the physical size of particles or species in a mixture or solution and is primarily used for water pre-treatment, treatment, and purification. There are almost 600 companies in worldwide, which offering membrane systems.

2. Electrospinning is a process used by the NF process, in which fibers are stretched and elongated down to a diameter of about 10 nm. The modified nanofibers that are produced are particularly useful in the filtration process as an ultra-concentrated filter with a very large surface area. Studies have found that electrospun nanofibers can capture metallic ions and are continually effective through re-filtration.

3. Surface modified membrane is a term used for membranes with altered makeup and configuration, though the basic properties of their underlying materials remain intact.

14.3 TYPES OF MEMBRANES

As it mentioned, membranes have achieved a momentous place in chemical technology and are used in a broad range of applications. The key property that is exploited is the ability of a membrane to control the permeation rate of a chemical species through the membrane. In essence, a membrane is nothing more than a discrete, thin interface that moderates the permeation of chemical species in contact with it. This interface may be molecularly homogeneous, that is completely uniform in composition and structure or it may be chemically or physically heterogeneous for example, containing holes or pores of finite dimensions or consisting of some form of layered structure. A normal filter meets this definition of a membrane, but, generally, the term filter is usually limited to structures that separate particulate suspensions larger than 1–10 μm [17].

The preparation of synthetic membranes is, however, a more recent invention, which has received a great audience due to its applications [18]. Membrane technology like most other methods has undergone a developmental stage, which has validated the technique as a cost-effective treatment option for water. The level of performance of the membrane technologies is still developing and it is stimulated by the use of additives to improve the mechanical and thermal properties, as well as the permeability, selectivity, rejection and fouling of the membranes [19]. Membranes can be fabricated to possess different morphologies. However, most membranes that have found practical use are mainly of asymmetric structure. Separation in membrane processes takes place as a result of differences in the transport rates of different species through the membrane structure, which is usually polymeric or ceramic [20].

The versatility of membrane filtration has allowed their use in many processes where their properties are suitable in the feed stream. Although membrane separation does not provide the ultimate solution to water treatment, it can be economically connected to conventional treatment technologies by modifying and improving certain properties [21].

The performance of any polymeric membrane in a given process is highly dependent on both the chemical structure of the matrix and the physical arrangement of the membrane [22]. Moreover, the structural impeccability of a membrane is very important since it determines its permeation and selectivity efficiency. As such, polymer membranes should be seen as much more than just sieving filters, but as intrinsic complex structures which can either be homogenous (isotropic) or heterogeneous (anisotropic), porous or dense, liquid or solid, organic or inorganic [22, 23].

14.3.1 ISOTROPIC MEMBRANES

Isotropic membranes are typically homogeneous/uniform in composition and structure. They are divided into three subgroups, namely: microporous, dense and electrically charged membranes [20]. Isotropic microporous membranes have evenly distributed pores (Figure 14.7a) [27]. Their pore diameters range between 0.01–10 μm and operate by the sieving mechanism. The microporous membranes are mainly prepared by the phase inversion method albeit other methods can be used.Conversely, isotropic dense membranes do not have pores and as a result they tend to be thicker than the microporous membranes (Figure 14.7b). Solutes are carried through the membrane by diffusion under a pressure, concentration or electrical potential gradient. Electrically charged membranes can either be

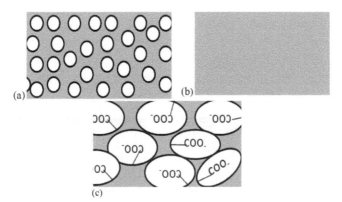

FIGURE 14.7 Schematic diagrams of isotropic membranes: (a) microporous; (b) dense; and (c) electrically charged membranes.

porous or non-porous. However in most cases they are finely microporous with pore walls containing charged ions (Figure 14.7c) [20, 28].

14.3.2 ANISOTROPIC MEMBRANES

Anisotropic membranes are often referred to as Loeb-Sourirajan, based on the scientists who first synthesized them [24, 25]. They are the most widely used membranes in industries. The transport rate of a species through a membrane is inversely proportional to the membrane thickness. The membrane should be as thin as possible due to high transport rates are eligible in membrane separation processes for economic reasons. Contractual film fabrication technology limits manufacture of mechanically strong, defect-free films to thicknesses of about 20 μm. The development of novel membrane fabrication techniques to produce anisotropic membrane structures is one of the major breakthroughs of membrane technology. Anisotropic membranes consist of an extremely thin surface layer supported on a much thicker, porous substructure. The surface layer and its substructure may be formed in a single operation or separately [17]. They are represented by non-uniform structures, which consist of a thin active skin layer and a highly porous support layer. The active layer enjoins the efficiency of the membrane, whereas the porous support layer influences the mechanical stability of the membrane. Anisotropic membranes can be classified into two groups, namely: (i) integrally skinned membranes where the active layer is formed from the same substance as the supporting layer, (ii) composite membranes where the polymer of the active layer differs from that of the supporting sub-layer [25]. In composite membranes, the layers are usually made from different polymers. The separation properties and permeation rates of the membrane are determined particularly by the surface

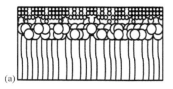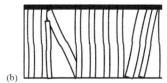

FIGURE 14.8 Schematic diagrams of anisotropic membranes: (a) Loeb-Sourirajan and (b) thin film composite membranes

layer and the substructure functions as a mechanical support. The advantages of the higher fluxes provided by anisotropic membranes are so great that almost all commercial processes use such membranes [17].

14.3.3 POROUS MEMBRANE

In Knudsen diffusion (Figure 14.9a), the pore size forces the penetrant molecules to collide more frequently with the pore wall than with other incisive species [26]. Except for some special applications as membrane reactors, Knudsen-selective membranes are not commercially attractive because of their low selectivity [27]. In surface diffusion mechanism (Figure 14.9b), the pervasive molecules adsorb on the surface of the pores so move from one site to another of lower concentration. Capillary condensation (Figure 14.9c) impresses the rate of diffusion across the membrane. It occurs when the pore size and the interactions of the penetrant with the pore walls induce penetrant condensation in the pore [28]. Molecular-sieve membranes in Figure 14.9d have gotten more attention because of their higher productivities and selectivity than solution-diffusion membranes. Molecular sieving membranes are means to polymeric membranes. They have ultra microporous (<7Å) with sufficiently small pores to barricade some molecules, while allowing others to pass through. Although they have several advantages such as permeation performance, chemical and thermal stability, they are still difficult to process because of some properties like fragile. Also they are expensive to fabricate.

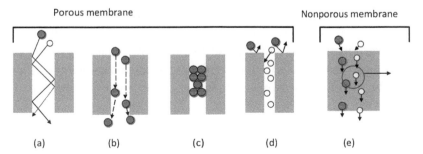

FIGURE 14.9   Schematic representation of membrane-based gas separations. (a) Knudsen-flow separation, (b) surface-diffusion, (c) capillary condensation, (d) molecular-sieving separation, and (e) solution-diffusion mechanism.

14.3.4   NONPOROUS (DENSE) MEMBRANE

Nonporous, dense membranes consist of a dense film through which permeants are transported by diffusion under the driving force of a pressure, concentration, or electrical potential gradient. The separation of various components of a mixture is related directly to their relative transport rate within the membrane, which is determined by their diffusivity and solubility in the membrane material. Thus, nonporous, dense membranes can separate permeants of similar size if the permeant concentrations in the membrane material differ substantially. Reverse osmosis membranes use dense membranes to perform the separation. Usually these membranes have an anisotropic structure to improve the flux [17].

The mechanism of separation by non-porous membranes is different from that by porous membranes. The transport through nonporous polymeric membranes is usually described by a solution–diffusion mechanism (Figure 14.9e). The most current commercial polymeric membranes operate according to the solution–diffusion mechanism. The solution–diffusion mechanism has three steps: (i) the absorption or adsorption at the upstream boundary, (ii) activated diffusion through the membrane, and (iii) desorption or evaporation on the other side. This solution–diffusion mechanism is driven by a difference in the thermodynamic activities existing at the upstream and downstream faces of the membrane as well as the intermolecular forces acting between the permeating molecules and those making up the membrane material.

The concentration gradient causes the diffusion in the direction of decreasing activity. Differences in the permeability in dense membranes are caused not only by diffusivity differences of the various species but also by differences in the physicochemical interactions of the species within the polymer. The solution–diffusion model assumes that the pressure within a membrane is uniform and that the chemical potential gradient across the membrane is expressed only as a concentration gradient. This mechanism controls permeation in polymeric membranes for separations.

14.4   CARBON NANOTUBES-POLYMER MEMBRANE

Iijima discovered carbon nanotubes (CNTs) in 1991 and it was really a revolution in nanoscience because of their distinguished properties. CNTs have the unique electrical properties and extremely high thermal conductivity [29, 30] and high elastic modulus (>1 TPa), large elastic strain – upto 5%, and large breaking strain – upto 20%. Their excellent mechanical properties could lead to many applications [31]. For example, with their amazing strength and stiffness, plus the advantage of lightness, perspective future applications of CNTs are in aerospace engineering and virtual biodevices [32].

CNTs have been studied worldwide by scientists and engineers since their discovery, but a robust, theoretically precise and efficient prediction of the mechanical properties of CNTs has not yet been found. The problem is, when the size of an object is small to nanoscale, their many physical properties cannot be modeled and analyzed by using constitutive laws from traditional continuum theories, since the complex atomistic processes affect the results of their macroscopic behavior. Atomistic simulations can give more precise modeled results of the underlying physical properties. Due to atomistic simulations of a whole CNT are computationally infeasible at present, a new atomistic and continuum mixing modeling method is needed to solve the problem, which requires crossing the length and time scales. The research here is to develop a proper technique of spanning multi-scales from atomic to macroscopic space, in which the constitutive laws are derived from empirical atomistic potentials which deal with individual interactions between single atoms at the micro-level, whereas Cosserat continuum theories are adopted for a shell model through the application of the Cauchy-Born rule to give the properties which represent the averaged behavior of large volumes of atoms at the macro-level [33, 34]. Since experiments of CNTs are relatively expensive at present, and often unexpected manual errors could be involved, it will be very helpful to have a mature theoretical method for the study of mechanical properties of CNTs. Thus, if this research is successful, it could also be a reference for the research of all sorts of research at the nanoscale, and the results can be of interest to aerospace, biomedical engineering [35].

Subsequent investigations have shown that CNTs integrate amazing rigid and tough properties, such as exceptionally high elastic properties, large elastic strain, and fracture strain sustaining capability, which seem inconsistent and impossible in the previous materials. CNTs are the strongest fibers known. The Young's Modulus of SWNT is around 1 TPa, which is 5 times greater than steel (200 GPa) while the density is only

1.2~1.4 g/cm^3. This means that materials made of nanotubes are lighter and more durable.

Beside their well-known extra-high mechanical properties, single-walled carbon nanotubes (SWNTs) offer either metallic or semiconductor characteristics based on the chiral structure of fullerene. They possess superior thermal and electrical properties so SWNTs are regarded as the most promising reinforcement material for the next generation of high performance structural and multifunctional composites, and evoke great interest in polymer based composites research. The SWNTs/polymer composites are theoretically predicted to have both exceptional mechanical and functional properties, which carbon fibers cannot offer [36].

14.4.1   CARBON NANOTUBES

Nanotubular materials are important "building blocks" of nanotechnology, in particular, the synthesis and applications of CNTs [37–39]. One application area has been the use of carbon nanotubes for molecular separations, owing to some of their unique properties. One such important property, extremely fast mass transport of molecules within carbon nanotubes associated with their low friction inner nanotube surfaces, has been demonstrated via computational and experimental studies [40, 41]. Furthermore, the behavior of adsorbate molecules in nano-confinement is fundamentally different than in the bulk phase, which could lead to the design of new sorbents [42].

Finally, their one-dimensional geometry could allow for alignment in desirable orientations for given separation devices to optimize the mass transport. Despite possessing such attractive properties, several intrinsic limitations of carbon nanotubes inhibit their application in large scale separation processes: the high cost of CNT synthesis and membrane formation (by microfabrication processes), as well as their lack of surface functionality, which significantly limits their molecular selectivity [43]. Although outer-surface modification of carbon nanotubes has been developed for nearly two decades, interior modification via covalent chemistry is still challenging due to the low reactivity of the inner-surface. Specifically, forming covalent bonds at inner walls of carbon nanotubes requires a transformation from sp^2 to sp^3 hybridization. The formation of sp^3 carbon is energetically unfavorable for concave surfaces [44].

Membrane is a potentially effective way to apply nanotubular materials in industrial-scale molecular transport and separation processes. Polymeric membranes are already prominent for separations applications due to their low fabrication and operation costs. However, the main challenge for using polymer membranes for future high-performance separations is to overcome the tradeoff between permeability and selectivity. A combination of the potentially high throughput and selectivity of nanotube materials with the process ability and mechanical strength of polymers may allow for the fabrication of scalable, high-performance membranes [45, 46].

14.4.2   STRUCTURE OF CARBON NANOTUBES

Two types of nanotubes exist in nature: multi-walled carbon nanotube) MWNTs), which were discovered by Iijima in 1991 [39] and SWNTs, which were discovered by Bethune et al. in 1993 [47, 48].

SWNT has only one single layer with diameters in the range of 0.6–1 nm and densities of 1.33–1.40 g/cm^3[49] MWNTs are simply composed of concentric SWNTs with an inner diameter is from 1.5 to 15 nm and the outer diameter is from 2.5 nm to 30 nm [50]. SWNTs have better defined shapes of cylinder than MWNT, thus MWNTs have more possibilities of structure defects and their nanostructure is less stable. Their specific mechanical and electronic properties make them useful for future high strength/modulus materials and nanodevices. They exhibit low density, large elastic limit without breaking (of up to 20–30% strain before failure), exceptional elastic stiffness, greater than 1000GPa and their extreme strength which is more than 20 times higher than a high-strength steel alloy. Besides, they also posses superior thermal and elastic properties: thermal stability up to 2800°C in vacuum and up to 750°C in air, thermal conductivity about twice as high as diamond, electric current carrying capacity 1000 times higher than copper wire [51]. The properties of CNTs strongly depend on the size and the chirality and dramatically change when SWCNTs or MWCNTs are considered [52].

CNTs are formed from pure carbon bonds. Pure carbons only have two covalent bonds: sp^2 and sp^3. The former constitutes graphite and the latter constitutes diamond. The sp^2 hybridization, composed of one s orbital and two p orbitals, is a strong bond within a plane but weak between planes. When more bonds come together, they form six-fold structures, like honey-comb pattern, which is a plane structure, the same structure as graphite [53].

Graphite is stacked layer by layer so it is only stable for one single sheet. Wrapping these layers into cylinders and joining the edges, a tube of graphite is formed, called nanotube [54].

Atomic structure of nanotubes can be described in terms of tube chiral-ity, or helicity, which is defined by the chiral vector, and the chiral angle, θ. Figure 14.10 shows visualized cutting a graphite sheet along the dotted lines and rolling the tube so that the tip of the chiral vector touches its tail. The chiral vector, often known as the roll-up vector, can be described by the following equation [55]:

$$C_h = na_1 + ma_2 \qquad (2)$$

As shown in Figure 14.10, the integers (n, m) are the number of steps along the carbon bonds of the hexagonal lattice. Chiral angle deter-mines the amount of "twist" in the tube. Two limiting cases exist where the chiral angle is at 0° and 30°. These limiting cases are referred to as zig-zag (0°) and armchair (30°), based on the geometry of the carbon bonds around the circumference of the nanotube. The difference in arm-chair and zig-zag nanotube structures is shown in Figure 14.11. In terms

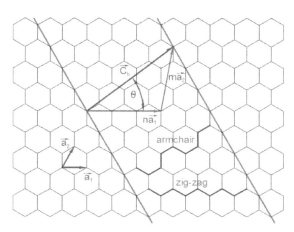

FIGURE 14.10   Schematic diagram showing how graphite sheet is 'rolled' to form CNT.

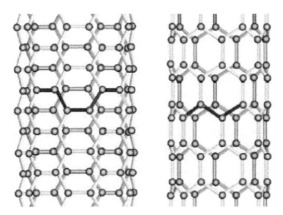

FIGURE 14.11 Illustrations of the atomic structure (a) an armchair and (b) a zig-zag nanotube.

of the roll-up vector, the zig-zag nanotube is (n, 0) and the armchair nanotube is (n, n). The roll-up vector of the nanotube also defines the nanotube diameter since the inter-atomic spacing of the carbon atoms is known.[36]

Chiral vector C_h is a vector that maps an atom of one end of the tube to the other. C_h can be an integer multiple a_1 of a_2, which are two basis vectors of the graphite cell. Then we have $C_h = a_1 + a_2$, with integer n and m, and the constructed CNT is called a (n,m) CNT, as shown in Figure 14.12. It can be proved that for armchair CNTs n=m, and for zigzag CNTs m=0. In Figure 14.12, the structure is designed to be a (4,0) zigzag SWCNT.

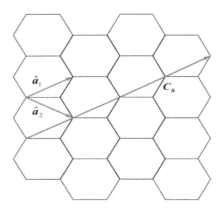

FIGURE 14.12 Basis vectors and chiral vector.

MWCNT can be considered as the structure of a bundle of concentric SWCNTs with different diameters. The length and diameter of MWCNTs are different from those of SWCNTs, which means, their properties differ significantly. MWCNTs can be modeled as a collection of SWCNTs, provided the interlayer interactions are modeled by Van der Waals forces in the simulation. A SWCNT can be modeled as a hollow cylinder by rolling a graphite sheet as presented in Figure 14.13.

If a planar graphite sheet is considered to be an undeformed configuration, and the SWCNT is defined as the current configuration, then the relationship between the SWCNT and the graphite sheet can be shown to be:

$$e_1 = G_1, e_2 = R\sin\frac{G_2}{R}, e_3 = R\cos\frac{G_2}{R} - R \tag{3}$$

The relationship between the integer's n, m and the radius of SWCNT is given by:

$$R = a\sqrt{m^2 + mn + n^2} / 2\pi \tag{4}$$

where a = $\sqrt{3}a_0$, and a_0 is the length of a non-stretched C-C bond which is 0.142 nm [56].

As a graphite sheet can be 'rolled' into a SWCNT, we can 'unroll' the SWCNT to a plane graphite sheet. Since a SWCNT can be considered as a rectangular strip of hexagonal graphite monolayer rolling up to a cylindrical tube, the general idea is that it can be modeled as a cylindrical shell,

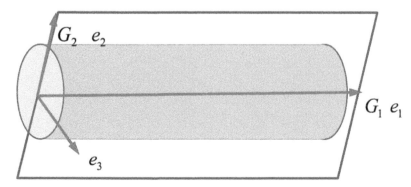

FIGURE 14.13   Illustration of a graphite sheet rolling to SWCNT.

a cylinder surface, or it can pull-back to be modeled as a plane sheet deforming into curved surface in three-dimensional space. A MWCNT can be modeled as a combination of a series of concentric SWCNTs with inter-layer inter-atomic reactions. Provided the continuum shell theory captures the deformation at the macro-level, the inner micro-structure can be described by finding the appropriate form of the potential function which is related to the position of the atoms at the atomistic level. Therefore, the SWCNT can be considered as a generalized continuum with microstructure [35].

14.4.3   CNT COMPOSITES

CNT composite materials cause significant development in nanoscience and nanotechnology. Their remarkable properties offer the potential for fabricating composites with substantially enhanced physical properties including conductivity, strength, elasticity, and toughness. Effective utilization of CNT in composite applications is dependent on the homogeneous distribution of CNTs throughout the matrix. Polymer-based nanocomposites are being developed for electronics applications such as thin-film capacitors in integrated circuits and solid polymer electrolytes for batteries. Research is being conducted throughout the world targeting the application of carbon nanotubes as materials for use in transistors, fuel cells, big TV screens, ultra-sensitive sensors, high-resolution atomic force microscopy (AFM) probes, super-capacitor, transparent conducting film, drug carrier, catalysts, and composite material. Nowadays, there are more reports on the fluid transport through porous CNTs/polymer membrane.

14.4.4   STRUCTURAL DEVELOPMENT IN POLYMER/CNT FIBERS

The inherent properties of CNT assume that the structure is well preserved (large-aspect-ratio and without defects). The first step toward effective reinforcement of polymers using nano-fillers is to achieve a uniform dispersion of the fillers within the hosting matrix, and this is also related to the as-synthesized nano-carbon structure. Secondly, effective interfacial interaction and stress transfer between CNT and polymer is essential for improved mechanical properties of the fiber composite.

Finally, similar to polymer molecules, the excellent intrinsic mechanical properties of CNT can be fully exploited only if an ideal uniaxial orientation is achieved. Therefore, during the fabrication of polymer/CNT fibers, four key areas need to be addressed and understood in order to successfully control the micro-structural development in these composites. These are: (i) CNT pristine structure, (ii) CNT dispersion, (iii) polymer–CNT interfacial interaction and (iv) orientation of the filler and matrix molecules (Figure 14.14). Figure 14.14 Four major factors affecting the micro-structural development in polymer/CNT composite fiber during processing [57].

Achieving homogenous dispersion of CNTs in the polymer matrix through strong interfacial interactions is crucial to the successful development of CNT/polymer nanocomposite [58]. As a result, various chemical or physical modifications can be applied to CNTs to improve its dispersion and compatibility with polymer matrix. Among these approaches acid treatment is considered most convenient, in which hydroxyl and carboxyl groups generated would concentrate on the ends of the CNT and at defect sites, making them more reactive and thus better dispersed [59, 60].

The incorporation of functionalized CNTs into composite membranes are mostly carried out on flat sheet membranes [61, 62]. For considering the potential influences of CNTs on the physicochemical properties of dope solution [63] and change of membrane formation route originated from various additives [64], it is necessary to study the effects of CNTs on the morphology and performance.

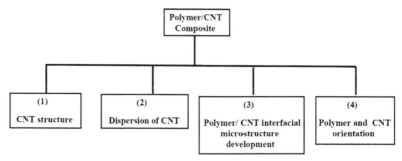

FIGURE 14.14 Four major factors affecting the micro-structural development in polymer/CNT composite fiber during processing.

14.4.5 GENERAL FABRICATION PROCEDURES FOR POLYMER/ CNT FIBERS

In general, when discussing polymer/CNT composites, two major classes come to mind. First, the CNT nano-fillers are dispersed within a polymer at a specified concentration, and the entire mixture is fabricated into a composite. Secondly, as grown CNT are processed into fibers or films, and this macroscopic CNT material is then embedded into a polymer matrix [65]. The four major fiber-spinning methods (Figure 14.15) used for polymer/CNT composites from both the solution and melt include dry-spinning [66], wet-spinning [67], dry-jet wet spinning (gel-spinning), and electrospinning [68]. An ancient solid-state spinning approach has been used for fabricating 100% CNT fibers from both forests and aero gels. Irrespective of the processing technique, in order to develop high-quality fibers many parameters need to be well controlled.

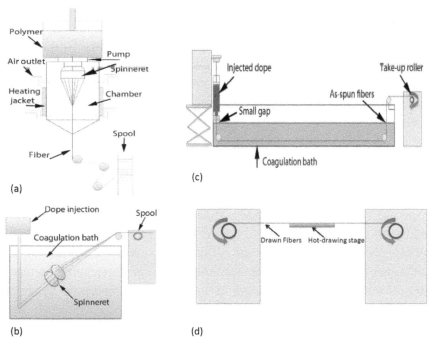

FIGURE 14.15 Schematics representing the various fiber processing methods (a) dry-spinning; (b) wet-spinning; (c) dry-jet wet or gel-spinning; and (d) post-processing by hot-stage drawing.

All spinning procedures generally involve: (i) Fiber formation (ii) coagulation/gelation/solidification, and (iii) drawing/alignment. For all of these processes, the even dispersion of the CNT within the polymer solution or melt is very important. However, in terms of achieving excellent axial mechanical properties, alignment and orientation of the polymer chains and the CNT in the composite is necessary. Fiber alignment is accomplished in post-processing such as drawing/annealing and is key to increasing crystallinity, tensile strength, and stiffness [69].

14.5   COMPUTATIONAL METHODS

Computational approaches to obtain solubility and diffusion coefficients of small molecules in polymers have focused primarily upon equilibrium molecular dynamics (MD) and Monte Carlo (MC) methods. These have been thoroughly reviewed by several investigators [70, 71].

Computational approach can play an important role in the development of the CNT-based composites by providing simulation results to help on the understanding, analysis and design of such nanocomposites. At the nanoscale, analytical models are difficult to establish or too complicated to solve, and tests are extremely difficult and expensive to conduct. Modeling and simulations of nanocomposites, on the other hand, can be achieved readily and cost effectively on even a desktop computer. Characterizing the mechanical properties of CNT-based composites is just one of the many important and urgent tasks that simulations can follow out [72].

Computer simulations on model systems have in recent years provided much valuable information on the thermodynamic, structural and transport properties of classical dense fluids. The success of these methods rests primarily on the fact that a model containing a relatively small number of particles is in general found to be sufficient to simulate the behavior of a macroscopic system. Two distinct techniques of computer simulation have been developed which are known as the method of molecular dynamics and the Monte Carlo method [73–75].

Instead of adopting a trial – and – error approach to membrane development, it is far more efficient to have a real understanding of the separation phenomena to guide membrane design [76–79]. Similarly, methods such as MC, MD and other computational techniques have improved the understanding of the relationships between membrane characteristics and separation properties. In addition to these inputs, it is also beneficial to have simple models and theories that give an overall insight into separation performance [80–83].

14.5.1 PRESENCE AND SELECTIVITY OF SEPARATION MEMBRANES

A membrane separates one component from another on the basis of size, shape or chemical affinity. Two characteristics dictate membrane performance, permeability, that is the flux of the membrane, and selectivity or the membrane's preference to pass one species and not another [84].

A membrane can be defined as a selective barrier between two phases, the "selective" being inherent to a membrane or a membrane processes. The membrane separation technology is proving to be one of the most significant unit operations. The technology inherits certain advantages over other methods. These advantages include compactness and light weight, low labor intensity, modular design that allows for easy expansion or operation at partial capacity, low maintenance, low energy requirements, low cost, and environmentally friendly operations. A schematic representation of a simple separation membrane process is shown in Figure 14.16.

A feed stream of mixed components enters a membrane unit where it is separated into a retentate and permeate stream. The retentate stream is typically the purified product stream and the permeate stream contains the waste component.

A quantitative measure of transport is the flux (or permeation rate), which is defined as the number of molecules that pass through a unit area

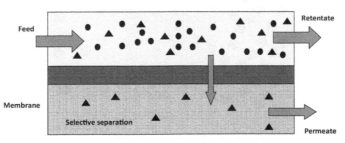

FIGURE 14.16   Schematic of membrane separation.

per unit time [85]. It is believed that this molecular flux follows Fick's first law. The flux is proportional to the concentration gradient through the membrane. There is a movement from regions of high concentration to regions of low concentration, which may be expressed in the form:

$$J = -D\frac{dc}{dx} \tag{5}$$

By assuming a linear concentration gradient across the membrane, the flux can be approximated as:

$$J = -D\frac{C_2 - C_1}{L} \tag{6}$$

where $C_1 = c(0)$ and $C_2 = c(L)$ are the downstream and upstream concentrations (corresponding to the pressures p_1 and p_2 via sorption isotherm $c(p)$, respectively, and L is the membrane thickness, as labeled in Figure 14.17.

The membrane performance of various materials is commonly compared using the thickness, independent material property, and the permeability, which is related to the flux as:

$$P = \frac{JL}{P_2 - P_1} = \left(\frac{C_2 - C_1}{P_2 - P_1}\right)D \tag{7}$$

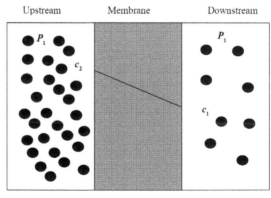

FIGURE 14.17 Separation membrane with a constant concentration gradient across membrane thickness L.

In the case where the upstream pressure is much greater than the downstream pressure ($p_2 \gg p_1$ and $C_2 \gg C_1$) the permeability can be simplified so:

$$P = \frac{C_2}{P_2} D \tag{8}$$

The permeability is more commonly used to describe the performance of a membrane than flux. This is because the permeability of a homogenous stable membrane material is constant regardless of the pressure differential or membrane thickness and hence it is easier to compare membranes made from different materials.

By introducing a solubility coefficient, the ratio of concentration over pressure C_2/p_2, when sorption isotherm can be represented by the Henry's law, the permeability coefficient may be expressed simply as:

$$P = SD \tag{9}$$

This form is useful as it facilitates the understanding of this physical property by representing it in terms of two components:

Solubility which is an equilibrium component describing the concentration of gas molecules within the membrane, that is the driving force, and Diffusivity, which is a dynamic component describing the mobility of the gas molecules within the membrane.

The separation of a mixture of molecules A and B is characterized by the selectivity or ideal separation factor $\alpha_{A/B} = P(A)/P(B)$, the ratio of permeability of the molecule A over the permeability of the molecule B. According to Eq. (9), it is possible to make separations by diffusivity selectivity $D(A)/D(B)$ or solubility selectivity $S(A)/S(B)$ [85, 86]. This formalism is known in membrane science as the solution – diffusion mechanism. Since the limiting stage of the mass transfer is overcoming of the diffusion energy barrier, this mechanism implies the activated diffusion. Because of this, the temperature dependences of the diffusion coefficients and permeability coefficients are described by the Arrhenius equations.

Gas molecules that encounter geometric constrictions experience an energy barrier such that sufficient kinetic energy of the diffusing molecule or the groups that form this barrier, in the membrane is required in order to overcome the barrier and make a successful diffusive jump. The common form of the Arrhenius dependence for the diffusion coefficient can be expressed as:

$$D_A = D_A^* \exp(-\Delta E_a / RT) \tag{10}$$

For the solubility coefficient the Van't Hoff equation holds:

$$S_A = S_A^* \exp(-\Delta H_a / RT) \tag{11}$$

where $\Delta H_a < 0$ is the enthalpy of sorption. From Eq. (9), it can be written:

$$P_A = P_A^* \exp(-\Delta E_P / RT) \tag{12}$$

where $\Delta E_p = \Delta E_a + \Delta H_a$ are known to diffuse within nonporous or porous membranes according to various transport mechanisms. Table 14.3 illustrates the mechanism of transport depending on the size of pores. For very

TABLE 14.3   Transport mechanisms

Mechanism	Schamatic	Process
Activated diffusion		Constriction energy barrier ΔE_a
Surface diffusion		Adsorption – site energy barrier ΔE_s
Knudsen diffusion		Direction and velocity $\overline{d}$ $\overline{U}$

narrow pores, size sieving mechanism is realized that can be considered as a case of activated diffusion. This mechanism of diffusion is most common in the case of extensively studied nonporous polymeric membranes. For wider pores, the surface diffusion (also an activated diffusion process) and the Knudsen diffusion are observed [87–89].

Sorption does not necessarily follow Henry's law. For a glassy polymer an assumption is made that there are small cavities in the polymer and the sorption at the cavities follows Langmuir's law. Then, the concentration in the membrane is given as the sum of Henry's law adsorption and Langmuir's law adsorption

$$C = K_P P + \frac{C_h^* b_P}{1 + b_P} \tag{13}$$

It should be noted that the applicability of solution (sorption)-diffusion model has nothing to do with the presence or absence of the pore.

14.5.2 DIFFUSIVITY

The diffusivity through membranes can be calculated using the time-lag method [90]. A plot of the flow through the membrane versus time reveals an initial transient permeation followed by steady state permeation. Extending the linear section of the plot back to the intersection of the x-axis gives the value of the time-lag (θ) as shown in Figure 14.18.

The time lag relates to the time it takes for the first molecules to travel through the membrane and is thus related to the diffusivity. The diffusion coefficient can be calculated from the time-lag and the membrane thickness as shown in Equation (14) [91–92].

$$D = \frac{\Delta x}{6\theta} \tag{14}$$

Surface diffusion is the diffusion mechanism, which dominates in the pore size region between activation diffusion and Knudsen diffusion [93].

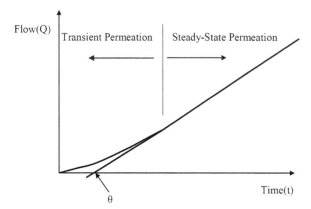

FIGURE 14.18   Calculation of the diffusion coefficient using the time-lag method, once the gradient is constant and steady state flow through the membrane has been reached, a extrapolation of the steady state flow line back to the x-axis where the flow is 0 reveals the value of the time lag (θ).

14.5.3   SURFACE DIFFUSION

A model that well described the surface diffusion on the pore walls was proposed many years ago. It was shown to be consistent with transport parameters in porous polymeric membranes. When the pore size decreased below a certain level, which depends on both membrane material and the permeability coefficient exceeds the value for free molecular flow (Knudsen diffusion), especially in the case of organic vapors. Note that surface diffusion usually occurs simultaneously with Knudsen diffusion but it is the dominant mechanism within a certain pore size. Since surface diffusion is also a form of activated diffusion, the energy barrier is the energy required for the molecule to jump from one adsorption site to another across the surface of the pore. By allowing the energy barrier to be proportionate to the enthalpy of adsorption, Gilliland et al. [94] established an equation for the surface diffusion coefficient expressed here as:

$$D_S = D_S^* \exp\left(\frac{-aq}{RT}\right) \tag{15}$$

where D_S^* is a pre-exponential factor depending on the frequency of vibration of the adsorbed molecule normal to the surface and the distance from one adsorption site to the next. The quantity the heat of adsorption is ($q > 0$) and a proportionality constant is ($0 < a < 1$). The energy barrier separates the adjacent adsorption sites. An important observation is that more strongly adsorbed molecules are less mobile than weakly adsorbed molecules [95].

In the case of surface diffusion, the concentration is well described by Henry's law $c = Kp$, where K is $K = K_0 \exp(q/RT)$ [95, 96]. Since solubility is the ratio of the equilibrium concentration over pressure, the solubility is equivalent to the Henry's law coefficient.

$$S_S = K_0 exp\left(q|RT\right) \tag{16}$$

Which implies the solubility is a decreasing function of temperature. The product of diffusivity and solubility gives:

$$P_S = P_S^* exp\left(\frac{(1-a)q}{RT}\right) \tag{17}$$

Since $0 < a < 1$ the total permeability will decrease with increased temperature meaning that any increase in the diffusivity is counteracted by a decrease in surface concentration [95].

14.5.4 KNUDSEN DIFFUSION

Knudsen diffusion [95, 97–99] depending on pressure and mean free path, which applies to pores between 10 Å and 500 Å in size [100]. In this region, the mean free path of molecules is much larger than the pore diameter. It is common to use Knudsen number $K_n = \lambda/d$ to characterize the regime of permeation through pores. When $K_n \ll 1$, viscous (Poiseuille) flow is realized. The condition for Knudsen diffusion is $K_n \gg 1$. An intermediate regime is realized when $K_n \approx 1$. The Knudsen diffusion coefficient can be expressed in the following form:

$$D_K = \frac{d}{3\tau}\bar{u} \tag{18}$$

This expression shows that the separation outcome should depend on the differences in molecular speed (or molecular mass). The average molecular speed is calculated using the Maxwell speed distribution as:

$$\bar{u} = \sqrt{\frac{8RT}{\pi m}} \tag{19}$$

And the diffusion coefficient can be presented as:

$$D_K = \left(\frac{d}{3\tau}\right)\left(\frac{8RT}{\pi m}\right)^{1/2} \tag{20}$$

For the flux in the Knudsen regime the following equation holds [101, 102]:

$$J = n\pi d^2 \Delta p D_K / 4RTL \tag{21}$$

After substituting Eq. (20) into Eq. (21), one has the following expressions for the flux and permeability coefficient is:

$$J = \left(\frac{n\pi^{\frac{1}{2}} d^3 \Delta p}{6\tau L}\right)\left(\frac{2}{mRT}\right)^{1/2} \tag{22}$$

$$P = \left(\frac{n\pi^{\frac{1}{2}} d^3}{6\tau}\right)\left(\frac{2}{mRT}\right)^{1/2} \tag{23}$$

Two important conclusions can be made from analysis of Eqs. (22) and (23). First, selectivity of separation in Knudsen regime is characterized by the ratio $\alpha_{ij} = (M_j/M_i)^{1/2}$. It means that membranes where Knudsen diffusion predominates are poorly selective.

The most common approach to obtain diffusion coefficients is equilibrium molecular dynamics. The diffusion coefficient that is obtained is a self-diffusion coefficient. Transport-related diffusion coefficients are less frequently studied by simulation but several approaches using non-equilibrium MD (NEMD) simulation can be used.

14.5.5   MOLECULAR DYNAMICS (MD) SIMULATIONS

Conducting experiments for material characterization of the nanocomposites is a very time consuming, expensive and difficult. Many researchers are now concentrating on developing both analytical and computational simulations. MD simulations are widely being used in modeling and solving problems based on quantum mechanics. Using Molecular dynamics it is possible to study the reactions, load transfer between atoms and molecules. If the objective of the simulation is to study the overall behavior of CNT-based composites and structures, such as deformations, load and heat transfer mechanisms then the continuum mechanics approach can be applied safely to study the problem effectively [103].

MD tracks the temporal evolution of a microscopic model system by integrating the equations of motion for all microscopic degrees of freedom. Numerical integration algorithms for initial value problems are used for this purpose, and their strengths and weaknesses have been discussed in simulation texts [104–106].

MD is a computational technique in which a time evolution of a set of interacting atoms is followed by integrating their equations of motion. The forces between atoms are due to the interactions with the other atoms. A trajectory is calculated in a 6-N dimensional phase space (three position and three momentum components for each of the N atoms). Typical MD simulations of CNT composites are performed on molecular systems containing up to tens of thousands of atoms and for simulation times up to nanoseconds. The physical quantities of the system are represented by averages over configurations distributed according to the chosen statistical ensemble. A trajectory obtained with MD provides such a set of configurations. Therefore the computation of a physical quantity is obtained as an arithmetic average of the instantaneous values. Statistical mechanics is the link between the nanometer behavior and thermodynamics. Thus the atomic system is expected to behave differently for different pressures and temperatures [107].

The interactions of the particular atom types are described by the total potential energy of the system, U, as a function of the positions of the individual atoms at a particular instant in time

$$U = U(X_i, \ldots, X_n) \tag{24}$$

where X_i represents the coordinates of atom i in a system of N atoms. The potential equation is invariant to the coordinate transformations, and is expressed in terms of the relative positions of the atoms with respect to each other, rather than from absolute coordinates[107].

MD is readily applicable to a wide range of models, with and without constraints. It has been extended from the original microcanonical ensemble formulation to a variety of statistical mechanical ensembles. It is flexible and valuable for extracting dynamical information. The Achilles' heel of MD is its high demand of computer time, as a result of which the longest times that can be simulated with MD fall short of the longest relaxation times of most real-life macromolecular systems by several orders of magnitude. This has two important consequences. (a) Equilibrating an atomistic model polymer system with MD alone is problematic; if one starts from an improbable configuration, the simulation will not have the time to depart significantly from that configuration and visit the regions of phase space that contribute most significantly to the properties. (b) Dynamical processes with characteristic times longer than approximately 10^{-7} s cannot be probed directly; the relevant correlation functions do not decay to zero within the simulation time and thus their long-time tails are inaccessible, unless some extrapolation is invoked based on their short-time behavior.

Recently, rigorous multiple time step algorithms have been invented, which can significantly augment the ratio of simulated time to CPU time. Such an algorithm is the reversible Reference System Propagator Algorithm (rRESPA) [108, 109]. This algorithm invokes a Trotter factorization of the Liouville operator in the numerical integration of the equations of motion: fast-varying (e.g., bond stretching and bond angle bending) forces are updated with a short time step Δt, while slowly varying forces (e.g., nonbonded interactions, which are typically expensive to calculate, are updated with a longer time step Δt. Using $\delta t = 1\,fs$ and, $\Delta t = 5ps$, one can simulate 300 ns of real time of a polyethylene melt on a modest workstation [110]. This is sufficient for the full relaxation of a system of C_{250} chains, but not of longer-chain systems.

A paper of Furukawa and Nitta is cited first to understand the NEMD simulation semi-quantitatively, since, even though the paper deals with various pore shapes, complicated simulation procedure is described clearly.

MD simulation is more preferable to study the non-equilibrium transport properties. Recently some NEMD methods have also been developed, such as the grand canonical molecular dynamics (GCMD) method [111, 112] and the dual control volume GCMD technique (DCV-GCMD) [113, 114]. These methods provide a valuable clue to insight into the transport and separation of fluids through a porous medium. The GCMD method has recently been used to investigate pressure-driven and chemical potential-driven gas transport through porous inorganic membrane [115].

14.5.5.1 Equilibrium MD Simulation

A self-diffusion coefficient can be obtained from the mean-square displacement (MSD) of one molecule by means of the Einstein equation in the form [115]:

$$D_A^* = \frac{1}{6N_\alpha} \lim_{t \to \infty} \frac{d}{dt} \left(r_i(t) - r_i(0) \right)^2 \tag{25}$$

where Na is the number of molecules, $r_i(t)$ and $r_i(0)$ are the initial and final (at time t) positions of the center of mass of one molecule i over the time interval t, and $(r_i(t) - r_i(0))^2$ is MSD averaged over the ensemble. The Einstein relationship assumes a random walk for the diffusing species. For slow diffusing species, anomalous diffusion is sometimes observed and is characterized by:

$$\left(r_i(t) - r_i(0) \right)^2 \propto t^n \tag{26}$$

where n < 1 (n = 1 for the Einstein diffusion regime). At very short times (t < 1 ps), the MSD may be quadratic iv n time (n = 2) which is characteristic of 'free flight' as may occur in a pore or solvent cage prior to collision with the pore or cage wall. The result of anomalous diffusion, which may or may not occur in intermediate time scales, is to create a smaller slope at short times, resulting in a larger value for the diffusion coefficient. At sufficiently long times (the hydrodynamic limit), a transition from anomalous to Einstein diffusion (n = 1) may be observed [71].

An alternative approach to MSD analysis makes use of the center-of-mass velocity autocorrelation function (VACF) or Green–Kubo relation, given as follows [116]

$$D = \frac{1}{3}\int (v_i(t).v_i(0))dt \qquad (27)$$

Concentration in the simulation cell is extremely low and its diffusion coefficient is an order of magnitude larger than that of the polymeric segments. Under these circumstances, the self diffusion and mutual diffusion coefficients of the penetrant are approximately equal, as related by the Darken equation in the following form:

$$D_{AB} = (D_A^* x_B + D_B^* x_A)\left(\frac{d\ln f_A}{d\,lnc_A}\right) \qquad (28)$$

In the limit of low concentration of diffusion $x_A \approx 0$, Equation (28) reduces to:

$$D_A^* \equiv D_{AB} \qquad (29)$$

14.5.5.2   Non-Equilibrium MD Simulation

Experimental diffusion coefficients, as obtained from time-lag measurements, report a transport diffusion coefficient, which cannot be obtained from equilibrium MD simulation. Comparisons made in the simulation literature are typically between time-lag diffusion coefficients (even calculated for glassy polymers without correction for dual-mode contributions and self-diffusion coefficients. As discussed above, mutual diffusion coefficients can be obtained directly from equilibrium MD simulation but simulation of transport diffusion coefficients require the use of NEMD methods, that are less commonly available and more computationally expensive [117].

For these reasons, they have not been frequently used. One successful approach is to simulate a chemical potential gradient and combine MD with GCMC methods (GCMC–MD), as developed by Heffelfinger and co-workers [114] and MacElroy [118]. This approach has been used to simulate permeation of a variety of small molecules through nanoporous carbon membranes, carbon nanotubes, porous silica and self-assembled monolayers [119–121]. A diffusion coefficient then can be obtained from the relation:

$$D = \frac{KT}{F}(V) \tag{30}$$

14.5.6   GRAND CANONICAL MONTE CARLO (GCMC) SIMULATION

A standard GCMC simulation is employed in the equilibrium study, while MD simulation is more preferable to study the non-equilibrium transport properties [104].

Monte Carlo method is formally defined by the following quote as: Numerical methods that are known as Monte Carlo methods can be loosely described as statistical simulation methods, where statistical simulation is defined in quite general terms to be any method that uses sequences of random numbers to perform the simulation [122].

The name "Monte Carlo" was chosen because of the extensive use of random numbers in the calculations [104]. One of the better known applications of Monte Carlo simulations consists of the evaluation of integrals by generating suitable random numbers that will fall within the area of integration. A simple example of how a MC simulation method is applied to evaluate the value of π is illustrated in Figure 14.19. By considering a square that inscribes a circle of a diameter R, one can deduce that the area of the square is R^2, and the circle has an area of $\pi R^2/4$. Thus, the relative area of the circle and the square will be $\pi/4$. A large number of two independent random numbers (with x and y coordinates) of trial shots is generated within the square to determine whether each of them falls inside of the circle or not. After thousands or millions of trial shots, the computer program keeps counting the total number of trial shots inside the square and the number of shots landing inside the circle. Finally, the value of $\pi/4$ can be approximated based on the ratio of the number of shots that fall inside the circle to the total number of trial shots.

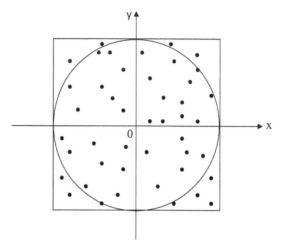

FIGURE 14.19 Illustration of the application of the Monte Carlo simulation method for the calculation of the value of π by generating a number of trial shots, in which the ratio of the number of shots inside the circle to the total number of trial shots will approximately approach the ratio of the area of the circle to the area of the square.

As stated earlier, the value of an integral can be calculated via MC methods by generating a large number of random points in the domain of that integral. Equation (31) shows a definite integral:

$$F = \int_a^b f(x)\, dx \tag{31}$$

where $f(x)$ is a continuous and real-valued function in the interval $[a, b]$. The integral can be rewritten as [104]:

$$F = \int_a^b dx \left(\frac{f(x)}{\rho(x)} \right) \rho(x) \cong \frac{f(\xi_i)}{\rho(\xi_i)}_\tau \tag{32}$$

If the probability function is chosen to be a continuous uniform distribution, then:

$$\rho(x) = \frac{1}{(b-a)} a \le x \le b \tag{33}$$

Subsequently, the integral, F, can be approximated as:

$$F \approx \frac{(b-a)}{\tau} \sum_{i=1}^{\tau} f(\xi_i) \tag{34}$$

In a similar way to the MC integration methods, MC molecular simulation methods rely on the fact that a physical system can be defined to possess a definite energy distribution function, which can be used to calculate thermodynamic properties.

The applications of MC are diverse such as Nuclear reactor simulation, Quantum chromo dynamics, Radiation cancer therapy, Traffic flow, Stellar evolution, Econometrics, Dow Jones forecasting, Oil well exploration, VSLI design [122].

The MC procedure requires the generation of a series of configurations of the particles of the model in a way, which ensures that the configurations are distributed in phase space according to some prescribed probability density.

The mean value of any configurational property determined from a sufficiently large number of configurations provides an estimate of the ensemble-average value of that quantity; the nature of the ensemble average depends upon the chosen probability density.

These machine calculations provide what is essentially exact information on the consequences of a given intermolecular force law. Application has been made to hard spheres and hard disks, to particles interacting through a Lennard-Jones 12–6 potential function and other continuous potentials of interest in the study of simple fluids, and to systems of charged particles [123].

The MC technique is a stochastic simulation method designed to generate a long sequence, or 'Markov chain' of configurations that asymptotically sample the probability density of an equilibrium ensemble of statistical mechanics [105, 116]. For example, a MC simulation in the canonical (NVT) ensemble, carried out under the macroscopic constraints of a prescribed number of molecules N, total volume V and temperature T, samples configurations r_p with probability proportional to $\exp[-\beta v(r_p)]$, with $\beta = 1/(k_B T)$, k_B being the Boltzmann constant and T the absolute temperature. Thermodynamic properties are computed as averages over all sampled configurations.

The efficiency of a MC algorithm depends on the elementary moves it employs to go from one configuration to the next in the sequence. An attempted move typically involves changing a small number of degrees of freedom; it is accepted or rejected according to selection criteria designed so that the sequence ultimately conforms to the probability distribution of interest. In addition to usual moves of molecule translation and rotation practiced for small-molecule fluids, special moves have been invented for polymers. The reptation (slithering snake) move for polymer chains involves deleting a terminal segment on one end of the chain and appending a terminal segment on the other end, with the newly created torsion angle being assigned a randomly chosen value [124].

In most MC algorithms the overall probability of transition from some state (configuration) m to some other state n, as dictated by both the attempt and the selection stages of the moves, equals the overall probability of transition from n to m; this is the principle of detailed balance or 'microscopic reversibility.' The probability of attempting a move from state m to state n may or may not be equal to that of attempting the inverse move from state n to state m. These probabilities of attempt are typically unequal in 'bias' MC algorithms, which incorporate information about the system energetics in attempting moves. In bias MC, detailed balance is ensured by appropriate design of the selection criterion, which must remove the bias inherent in the attempt [105, 116].

14.5.7   MEMBRANE MODEL AND SIMULATION BOX

The MD simulations [125] can be applied for the permeation of pure and mixed gases across carbon membranes with three different pore shapes: the diamond pore (DP), zigzag path (ZP) and straight path (SP), each composed of micro-graphite crystalline. Three different pore shapes can be considered: DP, ZP and SP.

Figure 14.20(a)–(c) shows the cross-sectional view of each pore shape. DP (a) has two different pore mouths; one a large (pore a) and the other a small mouth (pore b). ZP (b) has zigzag shaped pores whose sizes (diameters) are all the same at the pore entry. SP (c) has straight pores, which can be called slit-shaped pores.

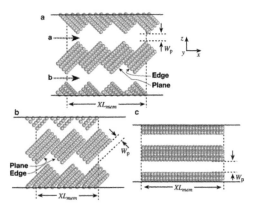

FIGURE 14.20 Three membrane pore shapes: (a) diamond path (DP), (b) zigzag path (ZP), (c) straight path (SP).

In a simulation system, we investigate the equilibrium selective adsorption and non-equilibrium transport and separation of gas mixture in the nanoporous carbon membrane are modeled as slits from the layer structure of graphite. A schematic representation of the system used in our simulations is shown in Figure 14.21(a) and (b), in which the origin of the coordinates is at the center of simulation box and transport takes place along the x-direction in the non-equilibrium simulations. In the equilibrium simulations, the box as shown in Figure 14.21(a) is employed, whose size is set as 85.20 nm × 4.92 nm × (1.675 + W) nm in x-, y-, and z-directions, respectively, where W is the pore width, that is, the separation distance between the centers of carbon atoms on the two layers forming a slit pore (Figure 14.21). L_{cc} is the separation distance between two centers of adjacent carbon atom; L_m is the pore length; W is the pore width, Δ is the separation distance between two carbon atom centers of two adjacent layers [126].

The simulation box is divided into three regions where the chemical potential for each component is the same. The middle region (M-region) represents the membrane with slit pores, in which the distances between the two adjacent carbon atoms (Lcc) and two adjacent graphite basal planes (Δ).

Period boundary conditions are employed in all three directions. In the non-equilibrium molecular dynamics simulations in order to use period boundary conditions in three directions, we have to divide the system into five regions as shown in Figure 14.21(b).

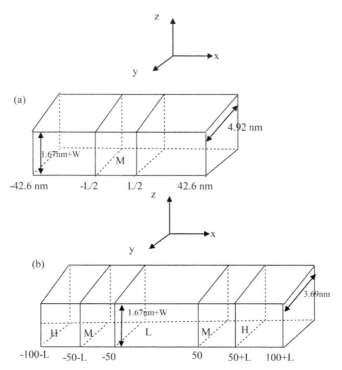

FIGURE 14.21 Schematic representation of the simulation boxes. The H-, L- and M-areas correspond to the high and low chemical potential control volumes, and membrane, respectively. Transport takes place along the x-direction in the non-equilibrium simulations. (a) Equilibrium adsorption simulations and (b) non-equilibrium transport simulations. L is the membrane thickness and W is the pore width.

Each symmetric box has three regions. Two are density control; H-region (high density) and L-region (low density) and one is free of control M-region, which is placed between the H- and L-region. For each simulation, the density in the H-region, ρ_H, is maintained to be that of the feed gas and the density in the L-region is maintained at zero, corresponding to the vacuum.

The difference in the gas density between the H- and L-region is the driving force for the gas permeation through the M-region, which represents the membrane.

The transition and rotational velocities are given to each inserted molecules randomly based on the Gaussian distribution around an average velocity corresponding to the specified temperature.

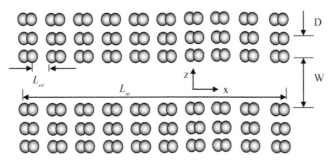

FIGURE 14.22   Schematic representation of slit pore.

Molecules spontaneously move from H- to L-region via leap-frog algorithm and a non-equilibrium steady state is obtained at the M-region. During a simulation run, equilibrium with the bulk mass at the feed side at the specified pressure and temperature is maintained at the H-region by carrying out GCMC creations and destructions in terms of the usual acceptance criteria [28]. Molecules entered the L-region were moved out immediately to keep vacuum. The velocities of newly inserted molecules were set to certain values in terms of the specified temperature by use of random numbers on the Gaussian distribution.

14.6   CONCLUDING REMARK

The concept of membrane processes is relatively simple but nevertheless often unknown. Membrane separation processes can be used for a wide range of applications. The separation mechanism in MF/UF/NF is mainly the size exclusion, which is indicated in the nominal ratings of the membranes. The other separation mechanism includes the electrostatic interactions between solutes and membranes, which depends on the surface and physiochemical properties of solutes and membranes. The available range of membrane materials includes polymeric, carbon, silica, zeolite and other ceramics, as well as composites. Each type of membrane can have a different porous structure. Nowadays, there are more reports on the fluid transport through porous CNTs/polymer membrane. Computational approach can play an important role in the development of the CNT-based composites by providing simulation results to help on the understanding, analysis and design of such nanocomposites. Computational approaches to obtain solubility and diffusion coefficients of small molecules in polymers have focused primarily upon molecular dynamics and Monte Carlo methods. Molecular dynamics simulations are widely being used in modeling and solving problems based on quantum mechanics. Using Molecular dynamics it is possible to study the reactions, load transfer between atoms and molecules. Monte Carlo molecular simulation methods rely on the fact that a physical system can be defined to possess a definite energy distribution function, which can be used to calculate thermodynamic properties. The Monte Carlo technique is a stochastic simulation method designed to generate a long sequence, or 'Markov chain' of configurations that asymptotically sample the probability density of an equilibrium ensemble of statistical mechanics. So using from molecular dynamic or Monte Carlo techniques can be useful to simulate the membrane separation process depends on the purpose and the condition of process.

KEYWORDS

- **computational methods**
- **filtration**
- **membrane**
- **membrane types**

REFERENCES

1. Majeed, S., et al., Multi-Walled Carbon Nanotubes (MWCNTs) Mixed Polyacrylonitrile (PAN) Ultrafiltration Membranes. Journal of Membrane Science, 2012, 403, 101–109.
2. Macedonio, F. E. Drioli, Pressure-Driven Membrane Operations and Membrane Distillation Technology Integration for Water Purification. Desalination, 2008, 223(1), 396–409.
3. Merdaw, A. A., Sharif, A. O., Derwish, G. A. W. Mass Transfer in Pressure-Driven Membrane Separation Processes, Part II. Chemical Engineering Journal, 2011, 168(1), 229–240.
4. Van Der Bruggen, B., et al., A Review of Pressure-Driven Membrane Processes in Wastewater Treatment and Drinking Water Production. Environmental Progress, 2003, 22(1), 46–56.
5. Cui, Z. F., Muralidhara, H. S. Membrane Technology: A Practical Guide to Membrane Technology and Applications in Food and Bioprocessing. 2010, Elsevier. 288.

6. Shirazi, S., Lin, C. J., Chen, D. Inorganic Fouling of Pressure-Driven Membrane Processes — A Critical Review. Desalination, 2010, 250(1), 236–248.
7. Pendergast, M. M., Hoek, E. M. V. A Review of Water Treatment Membrane Nanotechnologies. Energy and Environmental Science, 2011, 4(6), 1946–1971.
8. Hilal, N., et al., A comprehensive review of nanofiltration membranes: Treatment, pretreatment, modeling, and atomic force microscopy. Desalination, 2004, 170(3), 281–308.
9. Srivastava, A., Srivastava, S., Kalaga, K. Carbon Nanotube Membrane Filters, in Springer Handbook of Nanomaterials. 2013, Springer. 1099–1116.
10. Colombo, L. A. L. Fasolino, Computer-Based Modeling of Novel Carbon Systems and Their Properties: Beyond Nanotubes. Vol. 3. 2010, Springer. 258.
11. Polarz, S. B. Smarsly, Nanoporous Materials. Journal of Nanoscience and Nanotechnology, 2002, 2(6), 581–612.
12. Gray-Weale, A. A., et al., Transition-state theory model for the diffusion coefficients of small penetrants in glassy polymers. Macromolecules, 1997, 30(23), 7296–7306.
13. Rigby, D., Roe, R. Molecular Dynamics Simulation of Polymer Liquid and Glass. I. Glass Transition. The Journal of Chemical Physics, 1987, 87, 7285.
14. Freeman, B. D., Y. P. Yampolskii, I. Pinnau, Materials Science of Membranes for Gas and Vapor Separation. 2006, Wiley. com. 466.
15. Hofmann, D., et al., Molecular Modeling Investigation of Free Volume Distributions in Stiff Chain Polymers with Conventional and Ultrahigh Free Volume: Comparison Between Molecular Modeling and Positron Lifetime Studies. Macromolecules, 2003, 36(22), 8528–8538.
16. Greenfield, M. L., Theodorou, D. N. Geometric Analysis of Diffusion Pathways in Glassy and Melt Atactic Polypropylene. Macromolecules, 1993, 26(20), 5461–5472.
17. Baker, R. W., Membrane Technology and Applications. 2012, John Wiley & Sons. 592
18. Strathmann, H., L. Giorno, E. Drioli, Introduction to Membrane Science and Technology. 2011, Wiley-VCH Verlag & Company. 544.
19. Chen, J. P., et al., Membrane Separation: Basics and Applications, in Membrane and Desalination Technologies, L. K. Wang, et al., Editors. 2008, Humana Press. 271–332.
20. Mortazavi, S., Application of Membrane Separation Technology to Mitigation of Mine Effluent and Acidic Drainage. 2008, Natural Resources Canada. 194.
21. Porter, M. C., Handbook of Industrial Membrane Technology. 1990, Noyes Publications. 604.
22. Naylor, T. V., Polymer Membranes: Materials, Structures and Separation Performance. Rapra Technology Limited, 1996, 136.
23. Freeman, B. D., Introduction to Membrane Science and Technology. By Heinrich Strathmann. Angewandte Chemie International Edition, 2012, 51(38), 9485–9485.
24. Kim, I., H. Yoon, K. Lee, M. Formation of Integrally Skinned Asymmetric Polyetherimide Nanofiltration Membranes by Phase Inversion Process. Journal of Applied Polymer Science, 2002, 84(6), 1300–1307.
25. Khulbe, K. C., Feng, C. Y., Matsuura, T. Synthetic Polymeric Membranes: Characterization by Atomic Force Microscopy. 2007, Springer. 198.
26. Loeb, L. B., The Kinetic Theory of Gases. 2004, Courier Dover Publications. 678.
27. Koros, W. J., Fleming, G. K. Membrane-Based Gas Separation. Journal of Membrane Science, 1993, 83(1), 1–80.
28. Perry, J. D., Nagai, K., Koros, W. J. Polymer membranes for hydrogen separations. MRS bulletin, 2006, 31(10), 745–749.

29. Yang, W., et al., Carbon Nanotubes for Biological and Biomedical Applications. Nanotechnology, 2007, 18(41), 412001.
30. Bianco, A., et al., Biomedical Applications of Functionalised Carbon Nanotubes. Chemical Communications, 2005(5), 571–577.
31. Salvetat, J., et al., Mechanical Properties of Carbon Nanotubes. Applied Physics A, 1999, 69(3), 255–260.
32. Zhang, X., et al., Ultrastrong, Stiff, and Lightweight Carbon-Nanotube Fibers. Advanced Materials, 2007, 19(23), 4198–4201.
33. Arroyo, M., Belytschko, T. Finite Crystal Elasticity of Carbon Nanotubes Based on the Exponential Cauchy-Born Rule. Physical Review B, 2004, 69(11), 115415.
34. Wang, J., et al., Energy and Mechanical Properties of Single-Walled Carbon Nanotubes Predicted Using the Higher Order Cauchy-Born rule. Physical Review B, 2006, 73(11), 115428.
35. Zhang, Y., Single-walled carbon nanotube modeling based on one-and two-dimensional Cosserat continua. 2011, University of Nottingham.
36. Wang, S., Functionalization of Carbon Nanotubes: Characterization, Modeling and Composite Applications. 2006, Florida State University. 193.
37. Lau, K.-T., Gu, C., Hui, D. A critical review on nanotube and nanotube/nanoclay related polymer composite materials. Composites Part B: Engineering, 2006, 37(6), 425–436.
38. Choi, W., et al., Carbon Nanotube-Guided Thermopower Waves. Materials Today, 2010, 13(10), 22–33.
39. Iijima, S., Helical microtubules of graphitic carbon. nature, 1991, 354(6348), 56–58.
40. Sholl, D. S., Johnson, J. Making High-Flux Membranes with Carbon Nanotubes. Science, 2006, 312(5776), 1003–1004.
41. Zang, J., et al., Self-Diffusion of Water and Simple Alcohols in Single-Walled Aluminosilicate Nanotubes. ACS nano, 2009, 3(6), 1548–1556.
42. Talapatra, S., Krungleviciute, V., Migone, A. D. Higher Coverage Gas Adsorption on the Surface of Carbon Nanotubes: Evidence for a Possible New Phase in the Second Layer. Physical Review Letters, 2002, 89(24), 246106.
43. Pujari, S., et al., Orientation Dynamics in Multiwalled Carbon Nanotube Dispersions Under Shear Flow. The Journal of Chemical Physics, 2009, 130, 214903.
44. Singh, S., Kruse, P. Carbon Nanotube Surface Science. International Journal of Nanotechnology, 2008, 5(9), 900–929.
45. Baker, R. W., Future Directions of Membrane Gas Separation Technology. Industrial and Engineering Chemistry Research, 2002, 41(6), 1393–1411.
46. Erucar, I., Keskin, S. Screening Metal–Organic Framework-Based Mixed-Matrix Membranes for $CO_2/CH4$ Separations. Industrial and Engineering Chemistry Research, 2011, 50(22), 12606–12616.
47. Bethune, D. S., et al., Cobalt-Catalyzed Growth of Carbon Nanotubes with Single-Atomic-Layer Walls. Nature 1993, 363, 605–607.
48. Iijima, S., Ichihashi, T. Single-Shell Carbon Nanotubes of 1-nm Diameter. Nature, 1993, 363, 603–605.
49. Treacy, M., Ebbesen, T., Gibson, J. Exceptionally High Young's Modulus Observed for Individual Carbon Nanotubes. 1996.
50. Wong, E. W., Sheehan, P. E., Lieber, C. Nanobeam Mechanics: Elasticity, Strength, and Toughness of Nanorods and Nanotubes. Science, 1997, 277(5334), 1971–1975.

51. Thostenson, E. T., Li, C., Chou, T. W. Nanocomposites in Context. Composites Science and Technology, 2005, 65(3), 491–516.
52. Barski, M., Kędziora, P., Chwał, M. Carbon Nanotube/Polymer Nanocomposites: A Brief Modeling Overview. Key Engineering Materials, 2013, 542, 29–42.
53. Dresselhaus, M. S., Dresselhaus, G., Eklund, P. C. Science of Fullerenes and Carbon nanotubes: Their Properties and Applications. Academic Press, 1996, 965.
54. Yakobson, B., Smalley, R. E. Some Unusual New Molecules—Long, Hollow Fibers with Tantalizing Electronic and Mechanical Properties—have Joined Diamonds and Graphite in the Carbon Family. Am Scientist, 1997, 85, 324–337.
55. Guo, Y., Guo, W. Mechanical and Electrostatic Properties of Carbon Nanotubes under Tensile Loading and Electric Field. Journal of Physics D: Applied Physics, 2003, 36(7), 805.
56. Berger, C., et al., Electronic Confinement and Coherence in Patterned Epitaxial Graphene. Science, 2006, 312(5777), 1191–1196.
57. Song, K., et al., Structural Polymer-Based Carbon Nanotube Composite Fibers: Understanding the Processing–Structure–Performance Relationship. Materials, 2013, 6(6), 2543–2577.
58. Park, O. K., et al., Effect of Surface Treatment with Potassium Persulfate on Dispersion Stability of Multi-Walled Carbon Nanotubes. Materials Letters, 2010, 64(6), 718–721.
59. Banerjee, S., Hemraj-Benny, T., Wong, S. S. Covalent Surface Chemistry of Single-Walled Carbon Nanotubes. Advanced Materials, 2005, 17(1), 17–29.
60. Balasubramanian, K., Burghard, M. Chemically Functionalized Carbon Nanotubes. Small, 2005, 1(2), 180–192.
61. Xu, Z. L., Alsalhy Qusay, F. Polyethersulfone (PES) Hollow Fiber Ultrafiltration Membranes Prepared by PES/non-Solvent/NMP Solution. Journal of Membrane Science, 2004, 233(1–2), 101–111.
62. Chung, T. S., Qin, J. J., Gu, J. Effect of Shear Rate Within the Spinneret on Morphology, Separation Performance and Mechanical Properties of Ultrafiltration Polyethersulfone Hollow Fiber Membranes. Chemical Engineering Science, 2000, 55(6), 1077–1091.
63. Choi, J. H., Jegal, J., Kim, W. N. Modification of Performances of Various Membranes Using MWNTs as a Modifier. Macromolecular Symposia, 2007, 249–250(1), 610–617.
64. Wang, Z., Ma, J. The Role of Nonsolvent in-Diffusion Velocity in Determining Polymeric Membrane Morphology. Desalination, 2012, 286(0), 69–79.
65. Vilatela, J. J., Khare, R., Windle, A. H. The Hierarchical Structure and Properties of Multifunctional Carbon Nanotube Fibre Composites. Carbon, 2012, 50(3), 1227–1234.
66. Benavides, R. E., Jana, S. C., Reneker, D. H. Nanofibers from Scalable Gas Jet Process. ACS Macro Letters, 2012, 1(8), 1032–1036.
67. Gupta, V. B., Kothari, V. K. Manufactured Fiber Technology. 1997, Springer. 661.
68. Wang, T., Kumar, S. Electrospinning of Polyacrylonitrile Nanofibers. Journal of Applied Polymer Science, 2006, 102(2), 1023–1029.
69. Song, K., et al., Lubrication of Poly (vinyl alcohol) Chain Orientation by Carbon nano-chips in Composite Tapes. Journal of Applied Polymer Science, 2013, 127(4), 2977–2982.
70. Theodorou, D. N., Molecular Simulations of Sorption and Diffusion in Amorphous Polymers. Plastics Engineering-New York, 1996, 32, 67–142.

71. Müller-Plathe, F., Permeation of Polymers—A Computational Approach. Acta Polymerica, 1994, 45(4), 259–293.

72. Liu, Y. J., Chen, X. L. Evaluations of the Effective Material Properties of Carbon Nanotube-Based Composites Using a Nanoscale Representative Volume Element. Mechanics of Materials, 2003, 35(1), 69–81.

73. Gusev, A. A., Suter, U. W. Dynamics of Small Molecules in Dense Polymers Subject to Thermal Motion. The Journal of Chemical Physics, 1993, 99, 2228.

74. Elliott, J. A., Novel Approaches to Multiscale Modelling in Materials Science. International Materials Reviews, 2011, 56(4), 207–225.

75. Greenfield, M. L., Theodorou, D. N. Molecular Modeling of Methane Diffusion in Glassy Atactic Polypropylene via Multidimensional Transition State Theory. Macromolecules, 1998, 31(20), 7068–7090.

76. Peng, F., et al., Hybrid Organic-Inorganic Membrane: Solving the Tradeoff Between Permeability and Selectivity. Chemistry of materials, 2005, 17(26), 6790–6796.

77. Duke, M. C., et al., Exposing the Molecular Sieving Architecture of Amorphous Silica Using Positron Annihilation Spectroscopy. Advanced Functional Materials, 2008, 18(23), 3818–3826.

78. Hedstrom, J. A., et al., Pore Morphologies in Disordered NanoporousTthin Films. Langmuir, 2004, 20(5), 1535–1538.

79. Pujari, P. K., et al., Study of Pore Structure in Grafted Polymer Membranes Using Slow Positron Beam and Small-Angle X-ray Scattering Techniques. Nuclear Instruments and Methods in Physics Research Section B: Beam Interactions with Materials and Atoms, 2007, 254(2), 278–282.

80. Wang, X. Y., et al., Cavity Size Distributions in High Free Volume Glassy Polymers by Molecular Simulation. Polymer, 2004, 45(11), 3907–3912.

81. Skoulidas, A. I., Sholl, D. S. Self-Diffusion and Transport Diffusion of Light Gases in Metal-Organic Framework Materials Assessed Using Molecular Dynamics Simulations. The Journal of Physical Chemistry B, 2005, 109(33), 15760–15768.

82. Wang, X. Y., et al., A Molecular Simulation Study of Cavity Size Distributions and Diffusion in Para and Meta Isomers. Polymer, 2005, 46(21), 9155–9161.

83. Zhou, J., et al., Molecular Dynamics Simulation of Diffusion of Gases in Pure and Silica-Filled Poly (1-trimethylsilyl-1-propyne)[PTMSP]. Polymer, 2006, 47(14), 5206–5212.

84. Scholes, C. A., Kentish, S. E., Stevens, G. W. Carbon Dioxide Separation Through Polymeric Membrane Systems for Flue Gas Applications. Recent Patents on Chemical Engineering, 2008, 1(1), 52–66.

85. Wijmans, J. G., Baker, R. W. The Solution-Diffusion Model: A Unified Approach to Membrane Permeation. Materials Science of Membranes for Gas and Vapor Separation, 2006, 159–190.

86. Wijmans, J. G., Baker, R. W. The Solution-Diffusion Model: A Review. Journal of Membrane Science, 1995, 107(1), 1–21.

87. Way, J. D., Roberts, D. L. Hollow Fiber Inorganic Membranes for Gas Separations. Separation Science and Technology, 1992, 27(1), 29–41.

88. Rao, M. B., Sircar, S. Performance and Pore Characterization of Nanoporous Carbon Membranes for Gas Separation. Journal of Membrane Science, 1996, 110(1), 109–118.

89. Merkel, T. C., et al., Effect of Nanoparticles on Gas Sorption and Transport in Poly (1-trimethylsilyl-1-propyne). Macromolecules, 2003, 36(18), 6844–6855.

90. Mulder, M., Basic Principles of Membrane Technology Second Edition. 1996, Kluwer Academic Pub. 564.

91. Wang, K., Suda, H., Haraya, K. Permeation Time Lag and the Concentration Dependence of the Diffusion Coefficient of CO_2 in a Carbon Molecular Sieve Membrane. Industrial & Engineering Chemistry Research, 2001, 40(13), 2942–2946.

92. Webb, P. A., Orr, C. Analytical Methods in Fine Particle Technology. Vol. 55. 1997, Micromeritics Norcross, GA. 301.

93. Pinnau, I., et al., Long-Term Permeation Properties of Poly (1-trimethylsilyl-1-propyne) Membranes in Hydrocarbon—Vapor Environment. Journal of Polymer Science Part B: Polymer Physics, 1997, 35(10), 1483–1490.

94. Jean, Y. C., Characterizing Free Volumes and Holes in Polymers by Positron Annihilation Spectroscopy. Positron Spectroscopy of Solids, 1993, 1.

95. Hagiwara, K., et al., Studies on the Free Volume and the Volume Expansion Behavior of Amorphous Polymers. Radiation Physics and Chemistry, 2000, 58(5), 525–530.

96. Sugden, S., Molecular Volumes at Absolute Zero. Part II. Zero Volumes and Chemical Composition. Journal of the Chemical Society (Resumed), 1927, 1786–1798.

97. Dlubek, G., et al. Positron Annihilation: A Unique Method for Studying Polymers. in Macromolecular Symposia. 2004, Wiley Online Library.

98. Golemme, G., et al., NMR Study of Free Volume in Amorphous Perfluorinated Polymers: Comparsion with other Methods. Polymer, 2003, 44(17), 5039–5045.

99. Victor, J. G., Torkelson, J. M. On Measuring the Distribution of Local Free Volume in Glassy Polymers by Photochromic and Fluorescence Techniques. Macromolecules, 1987, 20(9), 2241–2250.

100. Royal, J. S., Torkelson, J. M. Photochromic and Fluorescent Probe Studies in Glassy Polymer Matrices. Macromolecules, 1992, 25(18), 4792–4796.

101. Yampolskii, Y. P., et al., Study of High Permeability Polymers by Means of the Spin Probe Technique. Polymer, 1999, 40(7), 1745–1752.

102. Kobayashi, Y., et al., Evaluation of Polymer Free Volume by Positron Annihilation and Gas Diffusivity Measurements. Polymer, 1994, 35(5), 925–928.

103. Huxtable, S. T., et al., Interfacial Heat Flow in Carbon Nanotube Suspensions. Nature materials, 2003, 2(11), 731–734.

104. Allen, M. P., Tildesley, D. J. Computer simulation of liquids. 1989, Oxford university press.

105. Frenkel, D., Smit, B., Ratner, M. A. Understanding molecular simulation: from algorithms to applications. Physics Today, 1997, 50, 66.

106. Rapaport, D. C., The art of Molecular Dynamics Simulation. 2004, Cambridge university press. 549.

107. Leach, A. R., Schomburg, D. Molecular modeling: principles and applications. 1996, Longman London.

108. Martyna, G. J., et al., Explicit Reversible Integrators for Extended Systems Dynamics. Molecular Physics, 1996, 87(5), 1117–1157.

109. Tuckerman, M., Berne, B. J., Martyna, G. J. Reversible Multiple Time Scale Molecular Dynamics. The Journal of Chemical Physics, 1992, 97(3), 1990.

110. Harmandaris, V. A., et al., Crossover from the Rouse to the Entangled Polymer Melt Regime: Signals from Long, Detailed Atomistic Molecular Dynamics Simulations, Supported by Rheological Experiments. Macromolecules, 2003, 36(4), 1376–1387.

111. Firouzi, M., Tsotsis, T. T., Sahimi, M. Nonequilibrium molecular dynamics simulations of transport and separation of supercritical fluid mixtures in nanoporous membranes. I. Results for a single carbon nanopore. The Journal of Chemical Physics, 2003, 119, 6810.

112. Shroll, R. M., Smith, D. E. Molecular Dynamics Simulations in the Grand Canonical Ensemble: Application to Clay Mineral Swelling. The Journal of Chemical Physics, 1999, 111, 9025.

113. Firouzi, M., et al., Molecular Dynamics Simulations of Transport and Separation of Carbon Dioxide–Alkane Mixtures in Carbon Nanopores. The Journal of Chemical Physics, 2004, 120, 8172.

114. Heffelfinger, G. S., van Swol, F. Diffusion in Lennard-Jones fluids using dual control volume grand canonical molecular dynamics simulation (DCV-GCMD). The Journal of Chemical Physics, 1994, 100, 7548.

115. Pant, P. K., Boyd, R. H. Simulation of Diffusion of Small-Molecule Penetrants in Polymers. Macromolecules, 1992, 25(1), 494–495.

116. Allen, M. P., Tildesley, D. J. Computer Simulation of Liquids. 1989, Oxford University Press. 385.

117. Cummings, P. T., Evans, D. J. Nonequilibrium Molecular Dynamics Approaches to Transport Properties and Non-Newtonian Fluid Rheology. Industrial and Engineering Chemistry Research, 1992, 31(5), 1237–1252.

118. MacElroy, J., Nonequilibrium Molecular Dynamics Simulation of Diffusion and Flow in Thin Microporous Membranes. The Journal of Chemical Physics, 1994, 101, 5274.

119. Furukawa, S. T. Nitta, Non-Equilibrium Molecular Dynamics Simulation Studies on Gas Permeation Across Carbon Membranes with Different Pore Shape Composed of Micro-Graphite Crystallites. Journal of Membrane Science, 2000, 178(1), 107–119.

120. Düren, T., Keil, F. J., Seaton, N. A. Composition Dependent Transport Diffusion Coefficients of CH_4/CF_4 Mixtures in Carbon Nanotubes by Non-Equilibrium Molecular Dynamics Simulations. Chemical Engineering Science, 2002, 57(8), 1343–1354.

121. Fried, J. R., Molecular Simulation of Gas and Vapour Transport in Highly Permeable Polymers. Materials Science of Membranes for Gas and Vapour Separation",, 2006, 95–136.

122. El Sheikh, A., Ajeeli, A., Abu-Taieh, E. Simulation and Modeling: Current Technologies and Applications. 2007, IGI Publishing.

123. McDonald, I., NpT-Ensemble Monte Carlo Calculations for Binary Liquid Mixtures. Molecular Physics, 2002, 100(1), 95–105.

124. Vacatello, M., et al., A Computer Model of Molecular Arrangement in a n-Paraffinic Liquid. The Journal of Chemical Physics, 1980, 73(1), 548–552.

125. Furukawa, S.-I., Nitta, T.Non-equilibrium molecular dynamics simulation studies on gas permeation across carbon membranes with different pore shape composed of micro-graphite crystallites. Journal of Membrane Science, 2000, 178(1), 107–119.

PART III

POLYMER CHEMISTRY

CHAPTER 15

STUDY OF DEGRADABLE POLYOLEFIN COMPOUNDS

T. I. AKSENOVA,[1] V. V. ANANYEV,[2] P. P. KULIKOV,[3] and
V. D. TRETYAKOVA[4]

[1]*Associated Professor, Moscow State University of Technologies and Management named after K.G. Razumovskiy, 73, Zemlyanoy Val St., Moscow, Russia; E-mail: aksentatyana@rambler.ru*

[2]*Head of Laboratory, Research Center of Ivan Fedorov Moscow State University of Printing Arts, 2A, Pryanishnikova St., Moscow, Russia; E-mail: vovan261147@rambler.ru*

[3]*Graduate Student, D. Mendeleyev University of Chemical Technology of Russia, 9, Miusskaya sq., Moscow, Russia; E-mail: p.kulikov.p@gmail.com*

[4]*Project Manager, METACLAY CJSC, 15, Karl Marx St., Karachev, Bryansk Region, Russia; E-mail: vera.d.tretyakova@gmail.com*

CONTENTS

ABSTRACT

The chapter discusses the effect manganese (degradation activator) organic complexes have on polyolefin compound biodegradabilities. Samples containing different quantities of the degradation activator were tested by two methods. The first one is testing by composting. The biodegradation process progress was judged by the samples' weight change, as well as by comparing the strain and strength characteristics data. The measurements were performed immediately after the preparation and after 30-, 60-, and 90-days conditioning. The second method is accelerated aging under the exposure to ultraviolet radiation. The study showed significant organic manganese complexes' effect on polyethylene and polypropylene biodegradability, which is of a greater level for polypropylene.

15.1   INTRODUCTION

The problem of environmental protection currently becomes global. In particular, a serious concern is caused by fast and practically uncontrolled growth of plastics consumption in many industries, resulting in sharp increase in wastes.

The most part of this amount belongs to foodstuff and consumer goods packing. Therefore, along with improvement in quality, reliability, and durability of packing materials improvement, a problem of their disposal upon expiration of the service life arises.

Polyolefins, especially polyethylene and polypropylene are most widely used in the packaging industry. At that, their natural degradation may last more than one hundred years. The existing recycling and disposal methods for such polymers are imperfect, for which reason a number of countries even restricted their production and consumption.

In the specialists' judgment, the "polymer waste" problem is to be solved by creating and mastering a range of polymers able to biodegrade under appropriate conditions without detrimental effects on the environment and human health [1]. It is biodegradation of polymer materials, which is today perhaps the most environmentally friendly way of packaging waste disposal [2]. Several fields of creation of such materials are currently under development, including:

- introduction of functional groups into the polymer structures, which help to accelerate photo- and oxydegradation [3];
- introduction of additives into large-tonnage polymers, which are able to initiate degradation of the base polymer to some extent.

The second way appears to be the most simple and relatively cheap method of solving the environmental problems. In this case, the materials obtained using the additives, as well as the additives themselves, shall be safe for the environment and human, which shall be supported by international certificates of conformance to the international regulations adopted in the area of composting and biodegradation (EN 13432, ASTM D 6400, Green PLA) [4].

As such, to modify polyolefins, an amphiphilic polymer was selected, with complexing groups, able to form stable complex compound with metal ions. As compared to salts of transition metals (Mn, Co, Cu, Ni, Fe) toxicity of polymer-metal complexes is significantly reduced which is indicated in a number of medical and pharmacological papers [5]. Owing to a favorable combination of physico-chemical properties of high-molecular-weight compounds and electrolytes, amphiphilic polymers gained strong position in many industrial, medical, scientific, and engineering fields. The use of polymer-metal complexes as degradation activators introduced into polymer films is reasonable due to their low toxicity, good compatibility with hydrophobic polymer matrix, and oxidative activity.

Therefore, the objective of this paper was to obtain and examine the properties, as well as degradability of polyolefin-based polymer compounds obtained through introduction of environmentally safe additives accelerating the polymer degradation process.

15.2 EXPERIMENTAL PART

The following were used as subjects of the research:

- Low-density polyethylene grade (LDPE) 15803–20;
- Polypropylene grade (PP) 21030–16.

The degradation activator is an additive with manganese organic complex.

To conduct the research, compounds with the contents shown in Table 15.1 had been prepared.

The lab extruder characteristics and temperature conditions of the compounds processing are presented in the Table 15.2.

Accelerated tests by exposure of the examined films to ultraviolet radiation were performed under the following conditions. The films were placed into a chamber isolated from external light sources. Two PRK-4 quartz lamps were used, providing radiation with a wavelength $\lambda = 185-315$ nm. The film samples as 100×100 mm squares were placed in 30 cm from ultraviolet lamps. It is known that 100-hour radiation in such a device is equivalent to approximately a year of the films' exposure to natural conditions.

The composting tests included exposure of the examined polymer material samples to microbially active soil (Fasco soil Spec. 0392–013–53910914–09 and Live Earth (TerraVita®) Spec. 0391–001–11158098–2002 were used).

The biodegradation process behaviors were judged by the weight change through weighing on analytical balance within 0.0001 g and the change in strain and strength characteristics of the materials after upon expiration of the selected soil exposure time. When performing the work, soil was used with a combined mineral fertilizer and moisture content of $60\pm5\%$ of its maximum water capacity. The total composting time was 30,

TABLE 15.1   Compound Contents

Compound	LDPE Concentration (%)	PP Concentration (%)	Additive Concentration (%)
LDPE (reference)	100	-	-
PP (reference)	-	100	-
Sample 1	99.5	-	0.5
Sample 2	99.0	-	1.0
Sample 3	98.5	-	1.5
Sample 4	-	99.5	0.5
Sample 5	-	99.0	1.0
Sample 6	-	98.5	1.5

TABLE 15.2   Lab Extruder Conditions

Lab Extruder Parameters	Values
Screw length (mm)	12
Effective screw length (length-to-diameter ratio), L/D	20
Maximum output (kg/h)	2.5
Temperature conditions for LDPE and compounds 1–3 processing:	
zone 1, °C	165
zone 2, °C	180
zone 3, °C	185
zone 4, °C	190
extrusion die, °C	195
Temperature conditions for PP and compounds 4–6 processing:	
zone 1, °C	190
zone 2, °C	235
zone 3, °C	255
zone 4, °C	260
extrusion die, °C	265

60, and 90 days. The polymer material samples and the reference sample were placed onto a soil substrate and fully covered with a soil layer, while providing constant access of air to the soil to avoid suppression of microorganisms' vital functions.

Tensile stress at break (σ) and elongation at break (ε) were determined at standard conditions onAI-7000-M testing machine equipped with a computer system for recording stress-strain curve. The work was performed in METACLAY CJSC Research Laboratory, Moscow.

15.3   RESULT AND DISCUSSION

The tests of the obtained compounds and original materials resulted in obtaining data allowing judgment on the effect the additive introduced has on degradability of polymer materials. The Figure 15.1 provides the strength test results for the samples, as a function of degradation activator content.

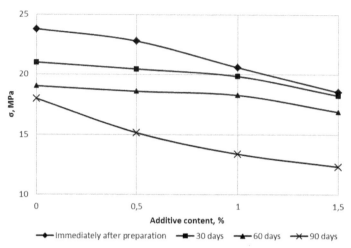

FIGURE 15.1   Relationship between LDPE film tensile failure stress and the additive content.

As the provided data show, increase in the additive content in a polyethylene compound leads do decrease in strength characteristics of both the original films and the films subjected to soil biodegradation. At that, the increase in additive concentration leads to a sharper decrease in the strength properties. Specifically, with 1.5 % additive after 90-days contact with soil, the film strength parameters decrease by two times.

It should be noted that similar results were observed for the strain characteristics. The measurements were performed both longitudinally and transversely.

The weight change in the samples was examined after the contact with soil for 30, 60, and 90 days. The weight of the samples containing 1.5 % additive increased by 6% after 90 days. This indicates formation of a more loose structure in the presence of manganese organic complex. At that, no significant weight changes were observed in LDPE containing 0%, 0.5%, 1% additive.

Similar examinations were performed with polypropylene-based compounds.

When a modifying additive was introduced, alteration of the films' strength characteristics was observed. The relationship between the PP film tensile failure stress and the complex content is shown in the Figure 15.2.

As seen on the diagram, increase in the compound's additive content up to 1,5 % results in increase of the strength parameters for PP films.

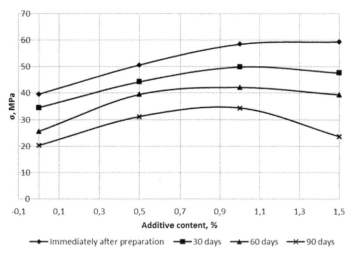

FIGURE 15.2   Relationship between the PP film tensile failure stress and the complex content.

At that, the polyethylene compounds showed a decrease in those parameters. This is possibly associated with the additive introduction providing plastification-orientational effects in PP.

However, after the PP samples' contact with soil, the same changes in the samples' properties were observed as with LDPE. The fastest acceleration of the samples' biodegradation in soil was demonstrated by compounds with 1.5% manganese organic complex.

Examining the effect of ultraviolet radiation on the samples produced from LDPE and PP compounds also allowed finding a significant difference in their behaviors.

Exposure of the LDPE compounds to ultraviolet radiation for 30 h did not result in alteration of the samples' properties – no changes in the film appearance were observed, the strength parameters of the original LDPE after the irradiation remained almost unaltered and the decrease in strength for the compounds containing 1 and 1.5% modifier did not exceed 15–20% of the initial values.

On the contrary, the PP compound samples containing degradation activator turned out to be very sensitive to UV radiation. The first changes in appearance of the original PP samples were observed after 30-hours exposure to UV, and for the compounds containing minimum (0.5%) activator, the changes in appearance were observed as early as after 10-hour irradiation, with complete failure of the samples in 30 h.

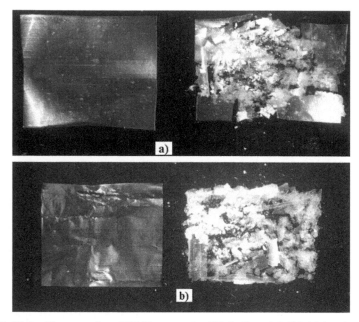

FIGURE 15.3   Photographs of PP film samples:0% (left) and 0.5% (right) degradation activator content: (a) after 10-hours UV irradiation; (b) after 30-hours UV irradiation.

The changes in film appearance resulted from UV irradiation are shown in the Figure 15.3.

The tests conducted allow concluding that the use of manganese organic complexes for creating biodegradable polymer compound is promising.

15.4   CONCLUSION

As a result of studying the effect manganese organic complex-based degradation activator has on biodegradation of polyolefins (LDPE and PP), the following conclusions were drawn:

1. Introduction of manganese organic complex into polyolefin-based compounds leads to both essential changes in their strain/strength characteristics, and change in the films' resistance to exposures: UV radiation and soil biodegradability.

2. Significant difference is observed in the changes caused by introduction of the degradation activator into LDPE and PP. For these

compounds, introduction of the additive leads to bidirectional alteration of strength characteristics.

3. Introduction of the degradation activator leads to essential acceleration of the compound biodegradation processes in the contact with soils, where increase in the additive level is accompanied with accelerated degradation.

4. Introduction of the additives into LDPE leads to insignificant changes in their properties upon UV irradiation. At that, the PP compounds containing even insignificant (0.5%) additives failed almost completely as early as after 10-hours exposure to UV.

The work was performed with financial support from the Ministry of Education and Science of the Russian Federation (RFMEFI60714X0002).

KEYWORDS

- **biodegradability**
- **degradation activator**
- **organic complexes**
- **polymer materials**
- **polyolefin compounds**

REFERENCES

1. Vo, T. H., Ivanova, T. V., Peshehonova, A. L., Sdobnikova, O. A., Samoylova, L. G., Kraus, S. V. Environmental aspects of polymer packaging materials disposal, 2008, p.62.
2. Stepanenko, A. B. A recycling alternative: biodegradable polymers. Waste recycling, 2006.
3. Zhang, Y., Guo, Sh., Lu, R., Gaofenzi Cailiao Kexue Yu Gongcheng. Polymer Mater Science Technology, 2003, p.14.
4. Rybas, S. V. Biodegradable polymers: an alternative to common plastics. Flexso Plus, 2010, p.40.
5. Kulikov, P. P., Goryachaya, A. V., Shtilman, M. I. Transition metal complexes with the amphiphilic poly N-vinylpyrrolidone. Biomaterials and Nanobiomaterials: Recent Problems and Safety Issues, 3rd Russian-Hellenic Symposium with International Participation and Young Scientists School, Iraklion, 2012, p.52.

CHAPTER 16

PRODUCTION OF SYNTHETIC RUBBER: A STUDY ON ECOLOGICAL SAFETY

R. R. USMANOVA[1] and G. E. ZAIKOV[2]

[1]*Ufa State Technical University of Aviation, 12 Karl Marks Str., Ufa 450100, Bashkortostan, Russia; E-mail: Usmanovarr@mail.ru*

[2]*N.M. Emanuel Institute of Biochemical Physics, Russian Academy of Sciences, 4 Kosygin Str., Moscow 119334, Russia; E-mail: chembio@sky.chph.ras.ru*

CONTENTS

ABSTRACT

Actions for decrease in gas exhausts of flares in synthetic rubber production are developed. The device is developed for wet purification of the gas exhausts, confirmed high degree of purification both in laboratory, and in industrial conditions. The packaging scheme of gas purifying for synthetic rubber production is implanted. The complex of the made research has formed the basis for designing of system of purification of air of industrial premises. Burning of gas exhausts on flares has allowed reducing pollution of air basin by toxic substances considerably.

16.1 INTRODUCTION

Hydrocarbons and their derivatives fall into the basic harmful exhausts of the petrochemical and oil refining enterprises. Actions for decrease in their pollutions are directed on elimination of losses of hydrocarbons at storage and transportation, and also on perfection of the control over hermetic sealing of the equipment and observance of a technological regime [1].

At many oil refining and petrochemical enterprises operate flares. They are intended for combustion formed at start-up of the equipment and in a process of manufacture of the gasses, which further processing, is economically inexpedient or impossible. To flares make following demands:

- the completeness of burning excluding formation of aldehydes, acids and harmful products;
- safe ignition, noiselessness and absence of a bright luminescence;
- absence of a smoke and carbon black; and
- stability of a torch at change of quantity and composition of gas exhausts.

Burning of gas exhausts on flares allows reducing pollution of air basin by toxic substances considerably. However, salvaging of waste gasses of the oil refining and petrochemical enterprises on flares is not a rational method of protection of environment. Therefore, it is necessary to provide decrease in exhausts of gasses on a torch. Application of effective systems of purification of gas exhausts result in to reduction of number of torches at the petrochemical enterprises [2].

Actions for protection of air basin yes the oil refining and petrochemical enterprises should be directed on increase of culture of production; strict observance of a technological mode; improvement of technology for the purpose of gas-making decrease; the maximum use of formed gasses; reduction of losses of hydrocarbons on objects of a manufacturing economy; working out and improvement of a quality monitoring and purification of pollutions.

16.2 ENGINEERING DESIGN AND EXPERIMENTAL RESEARCH OF NEW APPARATUSES FOR GAS CLEARING

Dynamic gas washer, according to Figures 16.1 and 16.2, contains the vertical cylindrical case with the bunker gathering slime, branch pipes of input and an output gas streams. Inside of the case it is installed conic vortex generator, containing. Dynamic gas washer works as follows [3].

The gas stream containing mechanical or gaseous impurity, acts on a tangential branch pipe in the ring space formed by the case and rotor. The liquid acts in the device by means of an axial branch pipe. At dispersion liquids, the zone of contact of phases increases and, hence, the effective utilization of working volume of the device takes place more.

FIGURE 16.1 Experimental installation "dynamic gas washer."

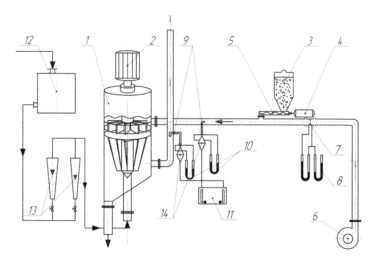

FIGURE 16.2 The circuit design of experimental installation: 1 – scrubber; 2 – the drive; 3 – the dust loading pocket; 4 – the electric motor; 5 – the batcher; 6 – the fan; 7 – a diaphragm; 8,10 – differential; 12 – the pressure tank; 13 – rotameter; 14 – sampling instruments.

The invention is directed on increase of efficiency of clearing of gas from mechanical and gaseous impurity due to more effective utilization of action of centrifugal forces and increase in a surface of contact of phases. The centrifugal forces arising at rotation of a rotor provide crushing a liquid on fine drops that causes intensive contact of gasses and caught particles to a liquid.

Owing to action of centrifugal forces, intensive hashing of gas and a liquid and presence of the big interphase surface of contact, there is an effective clearing of gas in a foamy layer. The aim was to determine the hydraulic resistance of irrigated unit when changing loads on the phases. The calculations take into account the angular velocity of rotation of the rotor blades and the direction of rotation of the swirl [4].

16.3   CLEARING OF GASES OF A DUST IN THE INDUSTRY

At the oil flares operate refining enterprise "synthetic rubber," in Bashkortostan. They are intended for combustion formed at start-up of the equipment and in a process of manufacture of gasses (*see*, Figure 16.3).

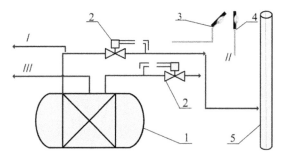

FIGURE 16.3   The schema of a flare with scrubber dynamic: 1 – scrubber dynamic; 2 – governor valves; 3 – the ignite burner; 4 – a pilot-light burner; 5 – a torch pipe; / – waste gas; // – fuel gas; /// – a condensate.

The dynamic scrubber is developed for decrease in gas exhausts of flares in synthetic rubber production.

Temperature of gasses of baking ovens in main flue gas a copper-recovery 500–600°C, after exhaust-heat boiler 250°C. An average chemical compound of smoke gasses (by volume): 17%CO_2; 16%N_2; 67% CO. Besides, in gas contains to 70 mg/m³ SO_2; 30 mg/m³ H_2S; 200 mg/m³ F and 20 mg/m³ CI. The gas dustiness on an exit from the converter reaches to 200/m³ the dust, as well as at a fume extraction with carbonic oxide after-burning, consists of the same components, but has the different maintenance of oxides of iron. In it than 1 micron, than in the dusty gas formed at after-burning of carbonic oxide contains less corpuscles a size less. It is possible to explain it to that at after-burning CO raises temperatures of gas and there is an additional excess in steam of oxides. Carbonic oxide before a gas heading on clearing burn in the special chamber. The dustiness of the cleared blast-furnace gas should be no more than 4 mg/m³. The following circuit design (*see*, Figure 16.4) is applied to clearing of the blast-furnace gas of a dust.

Gas from a furnace mouth of a baking oven 1 on gas pipes 3 and 4 is taken away in the gas-cleaning plant. In raiser and down taking duct gas is chilled, and the largest corpuscles of a dust, which in the form of sludge are trapped in the inertia sludge remover, are inferred from it. In a centrifugal scrubber 5 blast-furnace gas is cleared of a coarse dust to final dust content 5–10/m³ the Dust drained from the deduster loading pocket periodically from a feeding system of water or steam for dust moistening. The final

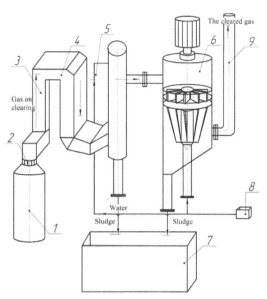

FIGURE 16.4 Process flow sheet of clearing of gas emissions: 1 – a Flare; 2 – water block; 3 – raiser; 4 – down-taking duct; 5 – centrifugal scrubber; 6 – scrubber dynamic; 7 – forecastle of gathering of sludge; 8 – hydraulic hitch; 9 – chimney.

cleaning of the blast-furnace gas is carried out in a dynamic spray scrubber where there is an integration of a finely divided dust. Most the coarse dust and drops of liquid are inferred from gas in the inertia mist eliminator. The cleared gas is taken away in a collecting channel of pure gas 9, whence is fed in an aerosphere. The clarified sludge from a gravitation filter is fed again on irrigation of apparatuses. The closed cycle of supply of an irrigation water to what in the capacity of irrigations the lime milk close on the physical and chemical properties to composition of dusty gas is applied is implemented. As a result of implementation of trial installation clearings of gas emissions the maximum dustiness of the gasses, which are thrown out in an aerosphere, has decreased with 3950 mg/m³ to 840 mg/m³, and total emissions of a dust from sources of limy manufacture were scaled down about 4800 to/a to 1300 to/a.

Such method gives the chance to make gas clearing in much smaller quantity, demands smaller capital and operational expenses, reduces an atmospheric pollution and allows to use water-recycling system [5].

TABLE 16.1   Results of Post-Test Examination

Compound	Concentration at the inlet, g/m^3	Concentration after clearing, g/m^3
Dust	0.02	0.00355
NO$_2$	0.10	0.024
SO$_2$	0.03	0.0005
CO	0.01	0.0019

16.4   CONCLUSIONS

1. The solution of an actual problem on perfection of complex system of clearing of gas emissions and working out of measures on decrease in a dustiness of air medium of the industrial factories for the purpose of betterment of hygienic and sanitary conditions of work and decrease in negative affecting of dust emissions given.
2. Designs on modernization of system of an aspiration of smoke gasses of a flare with use of the new scrubber which novelty is confirmed with the patent for the invention are devised. Efficiency of clearing of gas emissions is raised. Power inputs of spent processes of clearing of gas emissions at the expense of modernization of a flowchart of installation of clearing of gas emissions are lowered.
3. Ecological systems and the result of the recommendations implementation is to a high degree of purification of exhaust gasses and improve the ecological situation in the area of production. The economic effect of the introduction of up to 3 million rubles/year.

KEYWORDS

- dynamic scrubber
- flare
- purification of gas exhausts
- synthetic rubber

REFERENCES

1. Belov, P. S., Golubeva, I. A., Nizova, S. A. Production Ecology Chemicals from Petroleum Hydrocarbons and Gas, Moscow: Chemistry, 1991, 256 p.
2. Tetelmin, V. V., Jazev, V. A. Environment Protection in the Oil and Gas Complex, Dolgoprudnyj: Oil and Gas Engineering, 2009, 352 p.
3. Usmanova, R. R. Dynamic Gas Washer. Patent for the Invention of the Russian Federation №2339435. 20 November 2008. The Bulletin №33.
4. Shvydky, V. S., Ladygichev, M. G. Clearing of Gasses. The Directory, Moscow: Heat Power Engineering, 2002, 640 p.
5. Straus, V. Industrial Clearing of Gasses Moscow: Chemistry, 1981, 616 p.

CHAPTER 17

PROCESSES OF POLY(METHYL METHACRYLATE) DESTRUCTION BY HIGH-SPEED IMPACT AND PULSE LASER ACTION: A COMPARATIVE ANALYSIS

A. M. KUGOTOVA and B. I. KUNIZHEV

Kh.M. Berbekov Kabardino-Balkarian State University, Nalchik, Russia; E-mail: kam-02@mail.ru

CONTENTS

17.1   INTRODUCTION

The crater formation in and the destruction of poly(methyl methacrylate) (PMMA) under the high-speed impact are thoroughly studied in the Refs. [1, 2]. There is shown that craters in PMMA differ from those in metals not only in shape, but also in origin: the cavity in metals is formed in expense of a plastic flow while that in PMMA is a result of brittle fracture, formation of crack and ejection of matter in form of splinters. That is why in Ref. [1] the process of crater formation in PMMA is called the facial split which must by distinguished, by a number of criteria, from the rear spall resulting from the reflection of a shock wave from the back surface.

Such investigations continue in the Refs. [3, 4] where the analytical relations between the duration of penetration of the projectile into the target and the speed and crater depth have been established.

The laser action on PMMA is studied in Ref. [5] and the results are compared to those of the impact influence on target at the same energies. The evolution of the stressed state within the target and the processes of material's destruction have been studied and the damage zones and the degree of damage of the target material determined. The comparison of the results in both cases of influence leads the authors of Ref. [5] to the conclusion, that the laser action exerts more damage than the shock loading, because the destroyed zone occupies a larger volume.

The dependencies of the crater diameter D and of the depth of penetration of the polyethylene projectile into the PMMA target versus the speed of the projectile are presented in Table 17.1 (Experiments were conducted with help of the magneto-plasma accelerator of microparticles of the rail type [4]).

One can see from the Table 17.1 that despite the substantial differences between the pulsed laser action and the high-speed impact, the general pattern of the development of hydrodynamic processes in each case remains the same: the creation of the compression zone of the target material, the generation of a shock wave, the destruction of the target by waves reflected from the free surfaces. The identity of the occurring processes allowed

TABLE 17.1   The Dependencies of Crater Depth (h) and Diameter (D) in the Target from PMMA versus the Speed of Impact Loading

$\upsilon \times 10^3$, km/s	$h \times 10^{-3}$, m	$D \times 10^{-3}$, m	$E \times 10^{-3}$, J
0.80	4.1	12.1	0.52
1.25	5.5	12.5	1.27
1.75	6.2	13.2	2.49
2.20	8.0	14.3	3.97
3.0	12.0	14.4	7.38
3.7	17.8	14.6	11.23
4.8	18.5	14.8	13.12

the authors of that study to model the mentioned processes. The idea to use the energy of a laser impulse to studying of the high-speed impact is based on the assumption that the action of a laser pulse with energy E, duration τ and irradiated spot D_1 is similar to the effect of the projectile of diameter D, thickness L and velocity υ. This method of the study of a high-speed impact by means of a laser impulse allowed the authors of Ref. [5] to obtain a diagram of the stressed state in the target material, the position and size of the destroyed zones, the degree of damage in the material, as well as to assess the depth of penetration of the projectile into the target and to estimate the diameter of the crater in circumstances of the performed experiment. On the basis of obtained values of the energy of projectile, the characteristic parameters of the laser impulse were found from the previous study [1] according to the equation

$$\frac{m\upsilon^2}{2} = \alpha \cdot J \cdot \tau \cdot S \tag{1}$$

where m – mass of the projectile, υ – its speed, α – the absorption coefficient of the laser impulse, J – the density of power per irradiated surface, τ – the duration of laser impulse, S – the zone of the irradiated spot.

The dependencies of the dimensionless depth of penetration of the projectile made of PE (line 1) and of the laser impulse (lines 2 and 3) into the target h/D, calculated by us on the experimental data of studies [1, 5], are given in Figure 17.1. The projectile speed varied from 0.8 to 4.8 km/s in high-speed experiments while the energy of acting pulses ranged $(0.50 \div 13.2) \times 10^3$ J in both types of dynamic loading.

In both reports [1, 5], the target material (i.e., PMMA) was chosen such a way that it remained transparent after the intense dynamic loading, what allowed us to investigate the pattern of penetration of the projectile into the target in detail.

The authors of Ref. [6] showed that the dimensionless ratio h/D for craters formed in the targets by the collisions of macroparticles with velocities 2–10 km/s could be reduced, for different materials of the projectile, to the unique relation depending on the parameter $x=(\rho_y/\rho_\mu)\upsilon_0^2$, in km²/s². Thus we have the dependencies of the ratio h/D versus this variable x plotted in Figure 17.1.

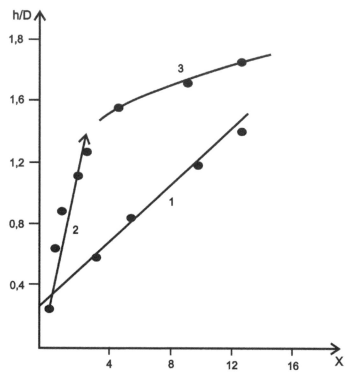

FIGURE 17.1   The dependence of the dimensionless depth of penetration of the projectile made of PE (line 1) and of the front of a laser impulse (lines 2 and 3) into the target made of PMMA.

Line 1 in the Figure 17.1 meets well with the equation

$$h/D = 0.86\, x_1. \tag{2}$$

It was obtained by the least squares method using the generalization of all the high-speed experiments conducted [1, 2] for the projectile made of PE and the target from PMMA.

Using the Eq. (1), we calculated h/D of the energy of laser impulses (Table 17.2). The lines 2 and 3 of the Figure 17.1 for the pulsed laser action have been obtained according to the Table 17.2. The lines 2 and 3 correspond, respectively, to the following equations:

$$h/D = 2.2\, x_2 \tag{3}$$

$$h/D = 0.45\, x_3 \tag{4}$$

TABLE 17.2 The Calculated Values of h/D per Energy of the Influence

$E \times 10^{-3}$, J	0.50	1.27	2.49	3.97	7.38	11.23	13.12
h/D (high-speed impact)	0.33	0.44	0.47	0.55	0.83	1.22	1.25
h/D (pulsed laser action)	0.62	0.88	1.16	1.25	1.57	1.60	1.72

The comparison of data presented in Fig. 17.1 and of the dependencies (2) and (3) demonstrates that the mechanisms of the brittle fracture of PMMA at high-speed impact and the pulsed laser action, with the energy of loading being the same, differ significantly. If one compares the coefficients at x of the straights (2)–(4) up to the energies of projectile $E=5\times10^{-3}$ J, then the coefficient at x_2 is 2.56 times greater than the coefficient at x_1, and 4.8 times in respect to that at x_3.

The mechanism of destruction of PMMA was thoroughly investigated in Ref. [4]. Based on the three-dimensional dependencies of the axial stress on the duration and depth of penetration, presented in that paper, one can describe the dependence of h/D versus E at high-speed impact both qualitatively and quantitatively. The analysis of data on high-dynamical destruction of the target made of PMMA given in Refs. [1, 4] and comparison of them to the data of Figure 17.1 (curve 1) reveals that at high speeds of loading of targets, firstly, the process of brittle fracture with the formation of the facial split takes place, then follows the increase of target temperature, thus the brittle fracture of the target transforms into the brittle-plastic destruction. Under the uniaxial compression, both longitudinal and transverse components of the stress increase. At the beginning of a high-speed loading, the changes of longitudinal and transverse stresses occur in consistent manner, then the damage threshold increases rapidly with the increase of the transverse compressive stress, and for some certain, critical, value the brittle-plastic transition takes place: the shear stresses become sufficient to activate the mechanisms of plastic deformation, and the disclosure of cracks is suppressed by the transverse strains.

The analysis of the lines (2) and (3) in Figure 17.2 reveals that the rate of deepening of penetration is almost three times greater at the beginning of the impact under the pulsed laser action compared to the high-speed impact, and then it becomes four times delayed (straight line). Further analysis of the pattern of the PMMA destruction by the laser impulse demonstrates that the channels formed by the laser action collapse as they move deeper into the target. The areas of intense plastic flow with a highly dispersed structure are formed near the channels. The formation of such structures cannot be explained by the concept suggested above, postulating elastic deformation and brittle fracture of the target by the advancing crack. The location of lines 2 and 3 in Figure 17.1 and the ratio of the coefficients at x_2 and x_3 correspond to the fact that at the very beginning (up to 6 microseconds) of the laser impulse begins and develops the brittle fracture, later transforming into the plastic failure of the target material in the zones located along the axis of the sample to the full depth of the target within one centimeter.

This mechanism of the destruction of PMMA, in the case of pulsed laser action, explains the absence of zones of axial and radial tensile stresses, which stretch to a depth of 0.5 cm from the front face of the target and also clarify the appearance of the rear spall, which was not found in the experiments and calculations of high-speed impacts [1, 2, 4].

17.2 CONCLUSION

From the discussed analysis, we can conclude that the use of laser impulses is more effective for creating the conditions for the spall fracture in comparison with the high-speed impact, although it has been assumed that the general scheme of the development of hydrodynamic processes remained the same in both cases [5]. Thus, we have shown in this study that the mechanisms of crater formation and of destruction in case of such pulse shock destructions differ in a great extent.

KEYWORDS

- comparative analysis
- destruction
- high-speed impact
- poly(methyl methacrylate)
- processes
- pulse laser action

REFERENCES

1. Kostin, V. V., Kunizhev, B. I., Suchkov, A. S., Temrokov, A. I. *ZhTF*. 1995, v.65. № 7, 176–179.
2. Pilyugin, N. N. *TVT*. 2004, v.42. № 3, 477–483.
3. Pilyugin, N. N., Ermolaev, I. K., Vinogradov, Y. A., Baulin N. N. *TVT*. 2002, v.40. № 5, 732–738.
4. Kugotova, A. M. Thesis PhD "High-Speed Loading and Destruction of Poly(Methyl Methacrylate)", Nalchik, 2009, p. 26.
5. Kostin, V. V., Kunizhev, B. I., Krasyuk, I. K., Kunizhev, B. I., Temrokov, A. I. *TVT*. 1997, v. 33. № 6, 962–967.
6. Titov, V. M., Fadeenko, Y. I. *Cosmic Investigations*. 1972, v.10, p.589.

CHAPTER 18

EFFECTIVE INHIBITORS OXIDATIVE DEGRADATION FOR POLYOLEFINS

M. M. MURZAKANOVA, T. A. BORUKAEV, and A. K. MIKITAEV

Kabardino-Balkarian State University, 173, Chernyshevsky Str., 360004, Nalchik, Russia; E-mail: m_m_murzakanova@mail.ru

CONTENTS

ABSTRACT

In this chapter, new effective inhibitors oxidative degradation of polymeric materials, are obtained in particular high-density polyethylene, which are not inferior to industrial known stabilizers.

18.1   INTRODUCTION

To save the basic physical and chemical properties of polymers in the process of their processing, storage and operation impose various stabilizers [1–5]. However, current stabilizers have a number of disadvantages: low thermal stability, migration, operate only on a particular mechanism, etc. In this connection there is a necessity of the search of new perspective stabilizers, deprived of the above-mentioned shortcomings.

18.2   EXPERIMENTAL PART

Taking into account these shortcomings, we have obtained new azomethine phenyl melamine antioxidants (AMPhMA) type, i.e., space-obstructed phenols, containing both azomethine and amino groups on the basis of 4-hydroxy-3,5-ditretbutilbenzaldegida and melamine, which have the following structure:

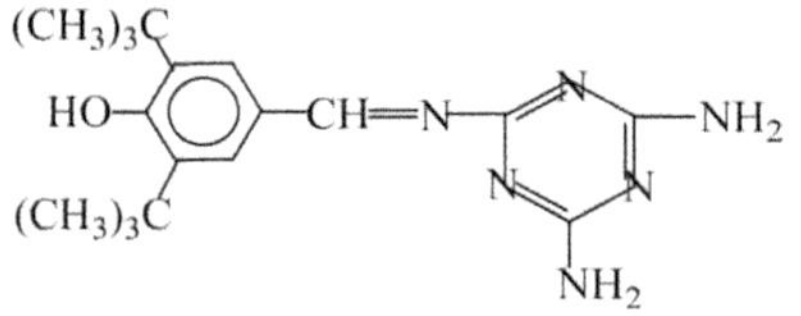

mono(-3,5-ditretbutyl-4-gidroksiphenilazometin)melamine – (monophenol)

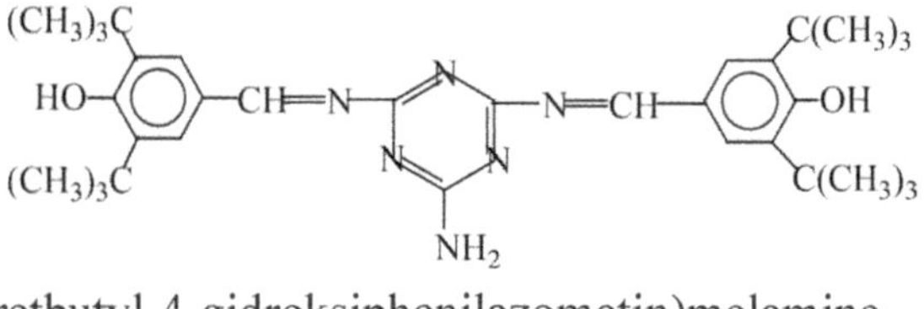

bis(-3,5-ditretbutyl-4-gidroksiphenilazometin)melamine – (bisphenol)

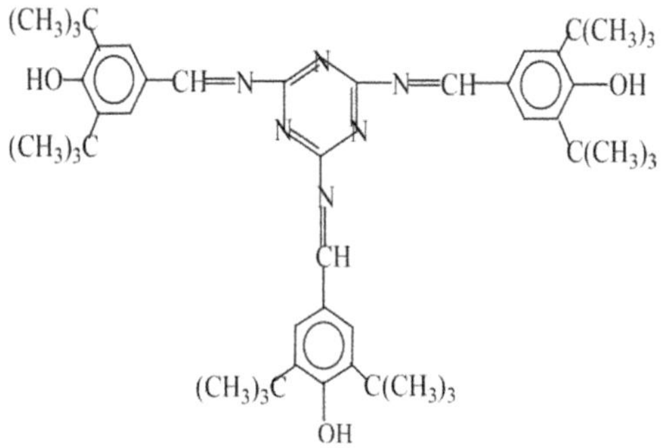

three(-3,5-ditretbutyl-4-gidroksiphenilazometin)melamine – (trifenol)

As can be seen from the chemical structure of molecules synthesized AMPhMA (monophenol, bisphenol and trifenol) have reactive centers, each of those can participate in radical processes [6–10]. In particular, the inhibiting ability of the obtained compounds may be due to the effect of pairing –CH=N– bond. The mechanism of action is following: PI – electrons of such systems are capable, when excited, moving to higher energy levels, which can accept free radicals. The acceptance of free radicals by CH=N will increase, especially at high temperatures.

The following capable for reactions center AMPhMA, which can inhibit radical processes, is an amino group. The mechanism of their action is in isolation of hydrogen from amino groups, which formed the radical that is stabilized by pairing it with p – electrons aromatic nucleus.

The third capable for reaction center AMPhMA is a hydroxyl group which, easily coming off the hydrogen atom, saturates the radicals produced during the degradation of polymers. The newly formed radical is, stabilized by shielding and p – electrons aromatic ring.

Thanks to these reactive centers, the obtained AMPhMA may be promising inhibitors of oxidative degradation of the polymer in particular polyolefins.

18.3 RESULTS AND DISCUSSIONS

To confirm the obtained above conclusion we have conducted the research of thermal properties of high density polyethylene (HDPE grade HDPE-276 and compositions HDPE + triphenol; HDPE + bisphenol and HDPE + monophenol. The received antioxidants were introduced in HDPE dispersing them in the melt at the stage of compounding. Antioxidant content was varied within 0.05–0.2% of mass.

One of the speedy and informative methods that allow for a comparative evaluation of thermal properties, stable and non-stable polymers, is thermogravimetric analysis (TGA) in the non-isothermal conditions [11]. Thermal characteristics HDPE and its compositions HDPE + AMPhMA were evaluated according to temperature of 2-, 5-, 10-, and 50% of loss of mass in the air. Comparative data TGA are given in Table 18.1.

Comparative analysis of the thermal properties of the source and compositions HDPE + AMPhMA showed the effectiveness of the compounds obtained as an inhibitor of oxidative degradation of HDPE. As can be seen from the Table 18.1, almost up to the content of 0.2% the azomethine phenyl melamine compounds the increase in the thermal properties of HDPE is observed. In its turn, this suggests that the obtained antioxidants are quite effective inhibitors of thermal-oxidative degradation. This follows from the comparison of the velocities of mass loss of samples in Table 18.3, where the decline of the mass for non-stabilized HDPE is happening faster and at lower temperatures than for stabilized samples.

In favor of the obtained conclusion of the betoken data TGA, the results of the research of thermostability of melts of HDPE and compositions HDPE + triphenol; HDPE + bisphenol and HDPE + monophenol.

TABLE 18.1 Results TGA Compositions on the Basis of HDPE

№	Sample	The Temperature of the Mass Loss on the Air, °C			
		2%	5%	10%	50%
1	HDPE (M-276)	235	310	370	410
2	HDPE (M-276) + 0.10% trifenol	250	315	372	415
3	HDPE (M-276) + 0.15% trifenol	255	330	375	420
4	HDPE (M-276) + 0.20% trifenol	268	380	400	428
5	HDPE (M-276) + 0.15% bisphenol	280	300	351	430
6	HDPE (M-276) + 0.2% bisphenol	275	300	356	425
7	HDPE (M-276) + 0.15% trifenol + 0.1% Fe/FeO	331	395	405	456
8	HDPE (M-276) + 0.10% Irganoks 1010	300	355	400	420

TABLE 18.2   Thermo Stability of the Melt and HDPE + Triphenol, HDPE + Bisphenol and HDPE+ Monophenol in the Process of Thermoxidation at 190°C

№	Sample	PTR, g/10 min	PTR/E$_r$°c			
			5 min	**10 min**	**20 min**	**30 min**
1.	HDPE (M-276)	0.71	0.88/24	1.05/48	1.37/93	1.17/65
2.	HDPE + 0.10% trifenol	0.48	0.40/–16	0.41/–14	0.18/–62	0.37/–23
3.	HDPE + 0.15% trifenol	0.47	0.43/–8	0.47/0	0.43/–8	0.39/–17
4.	HDPE + 0.20% trifenol	0.39	0.37/–5	0.38/–3	0.47/27	0.37/–5
5.	HDPE + 0.10% bisphenol	0.50	0.55/10	0.53/6	0.48/–4	0.55/10
6.	HDPE + 0.15% bisphenol	0.52	0.53/1.9	0.51/–1.9	0.47/–9.6	0.43/–17.3
7.	HDPE + 0.20% bisphenol	0.50	0.38/–24	0.38/–24	0.36/–28	0.48/–4
8.	HDPE + 0.10% monofenol	0.56	0.54/–3.6	0.59/5.4	0.56/0	0.51/–8.9
9.	HDPE + 0.15% monofenol	0.59	0.54/–8.4	0.57/–3.4	0.57/–3.4	0.50/–15.3
10.	HDPE + 0.20% monofenol	0.59	0.51/–13.6	0.51/–13.6	0.51/–13.6	0.42/–28.8
11.	HDPE + 0.10% melamin	0.62	0.69/11.3	1.23/93.3	1.27/105	0.48/25.8
12.	HDPE + 0.15% melamin	0.51	0.70/37.3	1.17/129	1.33/161	0.96/88.2
13.	HDPE + 0.20% melamin	0.62	0.81/30	1.47/137	1.46/135	1.10/77.4

Note: PTR is measured at 190°C and load of 2.16 kg.

Melt Thermostability of the initial and stabilized patterns (samples) was measured at the change of the values of melt flow index (BTI) depending on duration of thermal aging, at 190°C. It is known that the TPP subtly responds to chemical processes (destruction, structuring) in polymer melts [12].

The research results presented in Table 18.2 show a fairly high stabilizing efficiency of triphenol, bisphenol and monophenol in the EFF in the amount of upto 0.2% of mass.

TABLE 18.3 The Dependence of the Values of the TPP Source HDPE and Compositions HDPE + Triphenol; HDPE + Bisphenol and HDPE + Monophenol on Multiplicity Extrusion

№	Sample	PTR, g/10 min	PTR/ETS		
			The multiplicity of extrusion, n		
			1	2	3
1.	HDPE (M–276)	0.71	0.26/–63	0.47/–34	0.38/–46
2.	HDPE + 0.10% triphenol	0.48	0.36/–25	0.51/6	0.51/6
3.	HDPE + 0.15% triphenol	0.47	0.53/12.8	0.50/6	0.44/–6
4.	HDPE + 0.20% triphenol	0.39	0.47/–17	0.58/48	0.47/–17
5.	HDPE + 0.10% bisphenol	0.50	0.56/12	0.54/8	0.59/18
6.	HDPE + 0.15% bisphenol	0.52	0.49/–5.8	0.54/3.8	0.54\3.8
7.	HDPE + 0.20% bisphenol	0.50	0.57/14	0.60/7.1	0.61/22
8.	HDPE + 0.10% monophenol	0.56	0.61/8.9	0.62/10.7	0.65/16.1
9.	HDPE + 0.15% monophenol	0.59	0.70/18.6	0.63/6.8	0.65/10.2
10.	HDPE + 0.20% monophenol	0.59	0.62/5.1	0.57/–3.4	0.57/–3.4
11.	HDPE + 0.10% melamin	0.62	0.65/4.8	0.54/–12.9	0.56/–9.6
12.	HDPE + 0.15% melamin	0.51	0.47/–7.8	0.51/0	0.66/29.4
13	HDPE + 0.20% melamin	0.62	0.49/–20.9	0.48/–22.5	0.54/–12.9
14.	HDPE + 0.10% IRGANOKS1010	0.1010	0.081/–20	0.0825/–18	0.084/–17

Analysis of the data of Table 18.2 shows that samples of polymer containing the obtained stabilizers preserve original values of the TPP. Unregulated sample and the sample with melamine are weakly subjected to destruction after 5 min of exposure in the channel of PRT. Additionally, the Table 18.2 shows that the most optimal antioxidant content HDPE is 0.15 mass%.

Stabilizing properties of the synthesized compounds are confirmed by the results of testing of thermostability of melt multiple processing of melt HDPE and its composites. In particular testing of the efficiency of antioxidants as a "stabilizer processing," estimated the changes in the values of the BTI with multiple processing of melt carried out at extruder. The results of the research are presented in Table 19.3.

From Table 18.3, it follows that in the specified as AMPhMA when it's content in HDPE is more than 0.15 wt.% are quite effective. In particular samples HDPE, containing from 0.15 to 0.2% of monophenol, bisphenol and triphenol withstand three times processing, while the initial polymer and polymer with melamine after a single processing are structured (obviously filed). Therefore, it can be argued that the received stabilizers are quite effective stabilizers of processing and are not inferior and sometimes superior to industrial designs.

Thus, the results of thermal studies have shown that quite effective inhibitors oxidative degradation of polyolefin ace discovered.

It should be noted that the introduction into polymers different additives as a rule, leads to the changes in their basic physical and mechanical characteristics. It was interesting to investigate the influence of antioxidants synthesized on physical-mechanical properties of the polymer and determine the optimal content of the additives that does not result in deterioration of the original properties of the material. In this regard, the work for the assessment of the influence of additives-antioxidants on the physico-mechanical properties of HDPE the following characteristics were studied: Shore hardness scale (D) resiliency modulus (UOM) and tensile strength (EP), Izod impact strength (Ar). Test samples were prepared at the injection molding machine Politest firm Ray-Ran" at the temperature of a material cylinder 200°C temperature of form 60°C and locking pressure 8 ATM. The results of the measurements are shown in Table 18.4.

Measurement of hardness shore a scale D was conducted according to GOST 24621–91 on hard measuring instrument Hildebrand. For this purpose, we used samples of cylindrical shape with diameter of 40 mm and a height of 5 mm Maximum hardness and hardness after going through the process of relaxation were measured (Table 18.4).

Table 18.4 shows the introduction of azomethine phenyl melamine antioxidants in HDPE leads to the increase of their hardness on 2–4%. It is evidently connected with the fact that antioxidants, performing the role of a nucleating, increase the number of crystallization centers, contributing to the formation of fine-crystalline spherulite patterns. In its turn, this increases the hardness of the composite [13, 14]. These supramolecular transformations of polymer matrix will lead in its turn to the change of mechanical properties of the material: resilient modulus (UOM) and tensile strength (EP). So, impact strength (Ar) HDPE with the azomethine phenyl

TABLE 18.4 Physical and Mechanical Properties of HDPE, of the Stabilized by Azomethine Phenyl Melamine Antioxidants

№	Sample	Hardness shore a, D		A_p,* KJ/m^2	E_u, MPa	E_p, MPa	
		1 sek	15 sek			1 mm/min	10 mm/min
1.	(M-276)	67	58	21.1	656	898	1137
2.	HDPE + 0.10% monophenol	70	60	22.7	755	930	1159
3.	HDPE + 0.15% monophenol	67	60	26.3	773	990	1258
4.	HDPE + 0.20% monophenol	68	60	23.7	764	933	1186
5.	HDPE + 0.10% bisphenol	68	60	24.7	762	992	1251
6.	HDPE + 0.15% bisphenol	70	62	26.0	782	944	1231
7.	HDPE + 0.20% bisphenol	72	60	25.2	782	978	1222
8.	HDPE + 0.10% triphenol	70	60	29.8	676	957	1199
9.	HDPE + 0.15% triphenol	77	60	27.6	782	959	1221
10.	HDPE + 0.20% triphenol	71	60	29.1	769	950	1192

*Samples are tested with the cut of 0.5 mm.

melamine antioxidants, defined under GOST 19109-84 on the pendulum impact-testing machine company Gotech (Chinis) Ltd. in comparison with the initial polymer on 8–38%. This is due to the structural changes that occur in the original polymer at the introduction of the received antioxidants. In particular molecule antioxidants, occupying free volumes of amorphous phase polymer – HDPE, increase the dissipation of the energy of impact [14]. In its turn, tensile modulus of elasticity and bending of the compositions determined in accordance with GOST 9550-81-testing machine Gotech, are higher in comparison with the values of the original polymer. The increase of these values is probably connected with the location of particles of antioxidants in the free volume polymer amorphous phase, which leads to the change rigidity of the material.

18.4   CONCLUSION

Thus, new effective inhibitors oxidative degradation of polymeric materials, are obtained in particular high-density polyethylene, which are not inferior to industrial known stabilizers.

KEYWORDS

- **antioxidants**
- **polymers**
- **stabilizers**
- **thermo stability**

REFERENCES

1. Poods, V. S., Gromov, B. A., Perhaps Neumann, Sklyarov, E. Г. "Petroleum Chemistry". 1963, v.3. № 5. pp. 543.
2. Stevens, D. R., Guse, W. A. Patent USA. 2263582. 1941.
3. Hawkins, W. L., Lonza, V. L., Loeffler, B. B., Matreyer, W., Winslow, F. H. "Journal Applied Polymer Science". 1959, № 1. pp. 43.
4. Rubler Chemistry and Technology. 1959. № 4. pp. 1171.

5. Kalugina, E. V., Ivanov, A. N., Tochin, V. A. "Plastic compounds", 2006, №10. pp. 30.
6. Kalugina, E. V., Novotvorceva, T. N., Andreeva, M. B. Review P. M. 2001, № 6. pp. 29.
7. Berlin, A. A. "Achievements of Chemistry". 1975, 46. № 2. 44 pp.
8. Fly, Yu. B. "Plastic Compounds", 2007, №1. 15 pp.
9. Pospisil Jan, Nespurek Stanislav. Abstr. Mater. Res. Soc. Fall Meet., Boston, Mass. 995, p. 10.
10. Yendland, U. Thermal Methods of Analysis. M: World, 1978, 526 pp.
11. Kalinchev Ye. L., Sokovceva may BE, Properties and Processing of Thermoplastics. HP: Chemistry, 1983, 284 pp.
12. Hanim, H., Zarina, R., Ahmad Fuad, M. Y., Mohd. Ishak, Z. A., Azman Hassan. Malaesian Polimer Journal (MPJ). 2008, V. 3. № 12. pp 38–49.
13. Chan, C. M., Wu, J. S., Li, J. X., Cheung, Y. K. Polymer. 2002, v. 43. № 10. 2981–2992.
14. Prashatha, K., Soulestin, J., Lacrampe, M. F., Claes, M., Dupin, G., Krawczak, P. EXPREX Polimer Letters. 2008, v. 2. № 10.

CHAPTER 19

INVESTIGATION OF THE THERMAL DESTRUCTION PROCESS OF AROMATIC POLYAMIDE AND ORGANIC PLASTICS

A. I. BURYA, O. A. NABEREZHNAYA, and N. T. ARLAMOVA

Dneprodzerzhynsk State Technology University, Ukraine

CONTENTS

ABSTRACT

The methods for determining thermal resistance of polymer materials with optimization content of organic plastics based of aromatic polyamide phenylone C-25P are considered in this chapter, using the method of *Coats-Redfern*. This method determines the mechanism and kinetic parameters of the process thermal destruction of aromatic polyamide and organic plastics.

19.1 INTRODUCTION

Most often the behavior of materials under heating are characterized by thermal stability [1], the primary method of determining this characteristic is that of thermal gravimetric analysis (TGA). TGA analysis was applied to heat treated samples of aromatic polyamide phenylone C-25P (TU 6-05-221-101-71), which is a fine pink powder with a bulk density of 0.2–0.3 g/cm^3 and a specific viscosity of 0.5% solution in dimethylformamide with the amount of 5% lithium chloride not less than 0.75, and organic plastics (OP) based on it with different content of organic fibers: sulfone-T (possessing the following characteristics: fiber length of 3 mm; strength of 24–27 rkm; elongation at break – 17–21%; twist – 120–140 r/m) and phenylone.

Press-composition of the compound: C-25P phenylone +5–15 mass% of organic fibers (Table 19.1) was prepared by mixing the components in a rotating electromagnetic field in the presence of ferromagnetic particles. The thus prepared mixture of the product was processed into the block ware by compression molding at the pressure of 30 MPa and at the temperature of 598 K.

Thermal decomposition of the obtained samples was studied with the help of Q-1500D derivatographer of F. Paulik, J. Paulik and L. Erdey system, MOM Company (Hungary). The tests were carried out in special ceramic crucibles in air conditions within the temperature range of 298–873 K. The rate of temperature rise made 283 K/min; Al$_2$O$_3$ was used as a reference substance (inert), the substance addition made

TABLE 19.1 Compound Press Composition

Filler	Content, mass. %	Based	Content, mass. %
Fibers sulfon-T	–	Aromatic polyamide phenylone C-25P	100
	5		95
	10		90
	15		85
Fibers phenylon	5		95
	10		90
	15		85

100 mg. The sensitivity of DTG and DTA methods was that of 1/5 and 1/3, respectively.

The results are shown in Figure 19.1 and Table 19.2.

Table 19.2 shows that in case of introducing sulfone-T organic fiber to the composition of organic plastics, their thermostability is insignificantly improved. As for the organic fiber phenylon, the heat resistance of composites containing 15 mass% of the filler increases approximately by 1.1 times as compared with the initial polymer matrix.

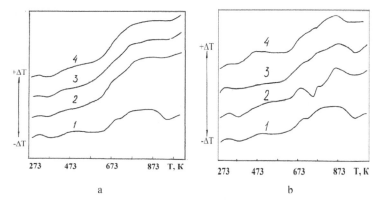

FIGURE 19.1 DTA curves for phenylon C-25P(1) and organic plastics based on it with different content of sulfone-T (a) and phenylon (b): 5(2); 10(3); 15(4) mass%.

TABLE 19.2 Thermo-Stability of Organic Plastics Based on Phenylon C-25P Containing Sulfone-T Fiber

Content fiber, mass. %	T_{10}	T_{20}	T_{30}	T_{50}	$T_{v\,max}$
Phenylone C-25P + organic fibers sulfone-T					
0	663	720	763	893	678
5	663	723	782	893	670
10	663	723	778	883	663
15	659	722	776	891	664
phenylone C-25P + organic fibers phenylone					
5	681	728	767	908	728
10	678	721	753	900	736
15	686	733	781	909	722

Note: T_{10}, T_{20}, T_{30}, T_{50} – temperature, K; 10, 20, 30, 50% weight loss.

Figure 19.2 shows that the contours of all the curves "weight loss – temperature" are similar, that is, the decomposition of the filled phenylon flows like that of pure one. In the first stage for all the investigated materials within the temperature range of 300–373 K a gradual decrease in mass (2–4%) has been observed due to the removal of moisture. Then, up to T = 613–623 K, the mass of the samples remained largely unchanged; while there is smooth running of DTA curves without explicit changes (Figure 19.1).

Intensive destruction of both initial phenylon and CM based on it accompanied by significant weight loss starts after the temperature reaches 673 K. DTA curves in this area show peaks relating to the decomposition of the material. Judging by the displacement of TG curves 2 and 3 relative to curve 1 (Figure 19.2b), the heat resistance of phenylon increases in case of introducing organic fiber phenylon.

In order to select the optimal kinetic model to describe the thermal degradation of phenylon C-25P and ETA based on it according to the experimental data obtained by TG and DTA analysis, we considered the possibility of using mathematical simulations of various heterogeneous processes [2].

The difficulties in assessing the kinetic parameters of the processes of the solids' thermal decomposition at a predetermined temperature range are known to be associated with a large number of conflicting data that has been obtained on the basis of kinetic equations describing various solid phase

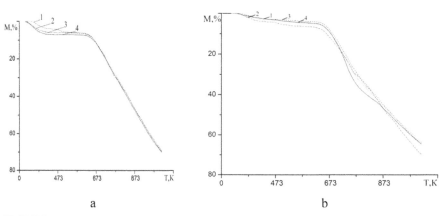

FIGURE 19.2   TG – curves of phenylon C-25P(1) of organic plastics based on based on it with different content of organic fibers: sulfone-T(a) and phenylone(b): 5(2); 10(3); 15(4) mass. %

conversions. However, the references consideration allows [3, 4] to state that there is a stereotype describing solid-phase processes with a selection of chemical- process conversion α as a criterion, which is determined by the formula [4]:

$$\alpha = (G_o - G)/(G_o - G_l) \tag{1}$$

where G_o, G, G_l is the initial, current and final mass of the sample.

The time dependence of the degree of conversion can be expressed by means of a differential equation [2]:

$$d\alpha/dt = k f(\alpha) \tag{2}$$

where τ is the time; k is the constant value of the reaction speed; $f(\alpha)$ is the algebraic function describing the mechanism of the process.

The dependency k on the temperature is described by means of a widely known Arrhenius equation [5]:

$$k = Z e^{-Eakt./RT} \tag{3}$$

where R is a universal gas constant, kJ/kg·K; e is the base of a natural logarithm; Z is a pre-exponential factor; $E_{акт.}$ is the apparent activation energy, kJ/mole.

Given the dependence (3), the Eq. (2) was presented as follows:

$$d\alpha/d\tau = Z e^{-Eakt./RT} f(\alpha) \tag{4}$$

Mathematically, for a non-isothermal kinetic analysis of the curve obtained with a linear heating, there are two methods: integral and differential ones. In this chapter, the method of Coates-Redfern [4] was used with applying the integral kinetic equations, which allows for consideration of non-isothermal reaction within the infinitesimal time interval as isothermal one. After integration and using the logarithms (4), the dependence looks as follows:

$$\lg \frac{k'(\alpha')}{T^2} = \lg \frac{ZR}{dT/d\tau \times E} \left(1 - \frac{2RT}{E}\right) - \left(\frac{E}{2,3RT}\right) \tag{5}$$

where

$$k`(\alpha) = \int_0^n \frac{dE}{df(\alpha)}; \quad k`(\alpha) = \frac{(1-\alpha)^{1-n} - 1}{n-1} \text{ for } n^1 \neq 1; \quad k'(\alpha) = -\ln(1-\alpha) \text{ for } n=1$$

(6)

Assuming that the relationship $[\lg k'(\alpha)] - [1/T]$ linear [3], in which case it can be used to determine the mechanism of heterogeneous reactions. This dependence is calculated directly from the experimental values α and T is linear only for such a function $k'(\alpha)$, which corresponds to the most probable processes controlling the actual speed of the reaction [5].

Determine the possible mechanism and calculation of the kinetic parameters of thermal destruction phenylon C-1 and CM based on it was carried

TABLE 19.3   Kinetic Equations of Different Mechanisms of Heterogeneous Processes [2]

Function	Equations	The process of determining the rate reaction	Mat. models
N_1	$k\tau = \alpha$	Nucleation by a power law $n = 1$	(7)
	$k\tau = 2\alpha^{1/2}$	——, ——, $n = 2$	(8)
R_2	$k\tau = 2[1 - (1-\alpha)^{1/2}]$	Reaction at the interface: - Cylindrical symmetry	(9)
R_3	$k\tau = 3[1 - (1-\alpha)^{1/3}]$	- Spherical symmetry	(10)
F_1	$k\tau = -\ln(1-\alpha)$	Random nucleation, one nucleus on each particle	(11)
A_2	$k\tau = 2[-\ln(1-\alpha)]^{1/2}$	Random nucleation uravnenie Avrahami-Erofeev, $n = 2$	(12)
A_3	$k\tau = 3[-\ln(1-\alpha)]$	——, ——, $n = 3$	(13)
A_4	$k\tau = 4[-\ln(1-\alpha)]^{1/4}$	——, ——, $n = 4$	(14)
D_1	$k\tau = 1/2\,\alpha^2$	One-dimensional diffusion	(15)
D_2	$k\tau = (1-\alpha)\ln(1-\alpha) + \alpha$	Two-dimensional diffusion, cylindrical symmetry	(16)
D_3	$k\tau = 3/2[1 - (1-\alpha^3)]^2$	Three-dimensional diffusion, spherical symmetry	(17)
D_4	$k\tau = \dfrac{3}{2}[(1-\dfrac{2}{3}\alpha) - (1-\alpha)^{2/3}]$	Two-dimensional diffusion equation Gistlinga-Brounshteyna	(18)

out using the integral kinetic equations of different mechanisms of heterogeneous processes (Table 19.3).

The criteria for selection of the mathematical simulation were direct correlation coefficient r within the coordinates of the Arrhenius equation and the minimum of S function:

$$S = f\{н\ a(\tau),\ T(\tau),\ \Delta T(\tau),\ E_{акт.},\ Z\}, \tag{19}$$

$$S = \sum_{i=1}^{n} \sqrt{\frac{\left(a_{э} - a_{p}\right)^2}{n}} \tag{20}$$

where $a_{э}$ and a_{p} are the experimental and calculated values of the degree of conversion, respectively; n is the number of experimental data; T is the temperature; E_{akt} is activation energy; Z is a pre-exponential factor.

Calculation results of the output parameters of phenylon C −25P thermal degradation, such as a correlation coefficient, the minimum of activation energy, and a pre-exponential factor, were all calculated by the program [4] developed for IBM and cited in Table 19.4.

TABLE 19.4   Kinetic Parameters of the Thermal Decomposition Process (Phenylon C-25P)

Mathematical models	r	S	$E_{akt.}$, kj/mol	lg Z
(7)	0.996	0.174×10^{-2}	33.96	−1.136
(8)	0.992	0.559×10^{-4}	11.99	−2.335
(9)	0.999	0.165×10^{-1}	47.86	−0.013
(10)	0.998	0.679×10^{-1}	53.35	0.43
(11)	0.994	0.139×10^{-1}	65.65	1.41
(12)	0.992	0.23×10^{-2}	27.84	−1.06
(13)	0.989	0.27×10^{-3}	15.23	−1.81
(14)	0.981	0.934×10^{-4}	8.9	−2.15
(15)	0.997	0.173×10^{-1}	77.89	1.56
(16)	0.999	0.245×10^{-2}	94.33	2.89
(17)	0.998	0.827×10^{-3}	116.66	4.21
(18)	0.999	0.307×10^{-2}	101.63	3.007

Thermal destruction of polymers which is very likely at a relatively high concentration of free radicals, in most cases has a radical chain mechanism [1] consisting of the following stages: I- initiation; II – chain development; III- chain transfer; IV – circuit failure.

High correlation coefficient values (Table 19.4) were obtained from the kinetic Eqs. (7), (9), (10), (16)–(18). Therefore, the minimum value of S was used as the main criterion for choosing the optimal mathematical model of the process. Thus, based on the data in Table 19.4, it was found that phenylon's thermal destruction process is best described by means of the reaction of the 1st order (7) equation and mathematical models (10), (18).

As it is known [6], during the thermal degradation of polymers containing aromatic rings in the chain, the main stage is that of chain initiation. This was confirmed by calculation, as the kinetic equation (10) describes the process of random nucleation: aromatic polyamide phenylon undergoes monomolecular conversion, which result in the formation of valence-saturated molecules' radicals possessing a relatively low reactivity.

Given that the thermolysis of phenylon primarily leads to the cleavage of the weakest Ph-N and C-N links [7], it can be assumed that the model (10) describes the below-mentioned homolytic process with the formation of free radicals.

$$(21)$$

Another way to adequately reflect the process is the mathematical model of the reactions at the interface (10). Obviously, the chain development occurs here as a result of heterogeneous reactions at the interface "polymer-gaseous thermolysis products" (CO_2, CO, H_2, H_2O, NH_3). In addition, the high correlation coefficient with a minimum value of S is observed in the case of two-dimensional diffusion (18) – cylindrical particles diffuse to the layer of ash accumulating as phenylon combusts. Obviously this is the slowest process, since it requires significant activation energy (Table 19.4). Similarly, the output parameters (r, S, E, $lg\,Z$) for the OP based on phenylon C-25P and containing fibers and phenylon were calculated. As expected,

TABLE 19.5   Kinetic Parameters of the Process of Organic Plastics' Thermal Decomposition

Mat. model	r	S	$E_{akt.}$, kJ/mol	lg Z
Phenylone C-25P +5 mass% of organic fibers sulfone-T				
(7)	0.916	0.342×10^{-3}	16.83	−2.12
(10)	0.909	0.155×10^{-1}	31.55	−0.76
(18)	0.936	0.84×10^{-2}	61.26	0.77
Phenylone C-25P +10 mass% of organic fibers sulfone-T				
(7)	0.935	0.735×10^{-3}	20.46	−1.86
(10)	0.919	0.385×10^{-1}	36.01	−0.45
(18)	0.943	0.389×10^{-2}	69.63	1.35
Phenylone C-25P +15 mass% of organic fibers sulfone-T				
(7)	0.871	0.235×10^{-3}	13.9	−2.36
(10)	0.879	0.113×10^{-1}	28.82	−0.98
(18)	0.914	0.335×10^{-3}	55.46	0.29
Phenylone C-25P +5 mass% of organic fibers phenylone				
(7)	0.921	0.461×10^{-2}	29.47	−1.25
(10)	0.915	0.119×10^{-2}	43.34	−0.03
(18)	0.933	0.525×10^{-2}	86.04	2.37
Phenylone C-25P +10 mass% of organic fibers phenylone				
(7)	0.967	0.872×10^{-1}	46.65	0.04
(10)	0.952	0.718×10^{-2}	66.32	1.71
(18)	0.964	0.749×10^{-2}	127.51	5.51
Phenylone C-25P +15 mass% of organic fibers phenylone				
(7)	0.996	0.701×10^{-2}	31.28	−1.17
(10)	0.904	0.217×10^{-2}	45.89	0.17
(18)	0.925	0.135×10^{-3}	90.54	2.72

the same mathematical model as in the case of the initial phenylo (7), (10), (18) can adequately describe ETA thermodestruction (Table 19.5).

Thus, it has been stated that the heat resistance of aromatic polyamide phenylon increases by 274–296 K in case of introducing organic fiber at the amount of 5–15 mass% (the most significant growth occurs in case of a 15% fill). As for the organic fiber sulfone-T, the thermal resistance increases insignificantly by 277–292 K.

KEYWORDS

- analysis
- aromatic polyamides
- organic plastics
- polymers
- thermal destruction
- thermal resistance

REFERENCES

1. Korshak, V. V. Chemical structure and thermal characteristics of polymers. Moscow: "Nauka" ("Science", in Rus.) Publishing House, 1970, p. 367.
2. Sestak, J. Theory of thermal analysis: "Mir" ("Peace", in Rus.) Publishing House, 1987, p. 456.
3. Zuru, A. A., Whitehead, R., Criffiths, D. L. A new technique for determination of the possible reaction mechanism from non-isothermal thermogravimetric data. Thermochim. Acta, 164, 305, 1990, R.285.
4. Bashtannik, P. I., Sitnic, S. V. Kinetic analysis of thermogravimetric data thermoplastic matrices polyacetal. Mechanics of Composite materials. 1994, N. 6, 843–847.
5. Emanuel, N. M., Knorre, D. G. The Rate of Chemical Kinetics. Moscow: "Higher School" Publishing House (in Rus.), 1969, p. 432.
6. Sokolov, L. B. Heat-resistant and high-strength polymeric material. Moscow: "Knowledge" Publishing House (in Rus.), 1984, p. 64.
7. Sokolov, L. B., Gerasimov, V. D., Savinov, V. D., Belyakov, V. K. Resistant aromatic polyamides. Moscow: "Khimiya" ("Chemistry", in Rus.) Publishing House, 1975. p. 256.

CHAPTER 20

METHACRYLATE GUANIDINE AND METHACRYLOYL GUANIDINE HYDROCHLORIDE: SYNTHESIS AND POLYMERIZATION

M. R. MENYASHEV, A. I. MARTYNENKO, N. I. POPOVA, N. A. KLESHCHEVA, and N. A. SIVOV

A.V. Topchiev Institute of Petrochemical Synthesis, RAS (TIPS RAS), Moscow, Russia

CONTENTS

ABSTRACT

We have found an efficient way of guanidine methacrylate synthesis (single-stage). It were studied methods of synthesis (meth)acryloyl guanidine by reacting guanidine with an acid chlorides, methyl(meth)acrylates, and appropriate acids. Polymerization of guanidine methacrylate and methacryloyl guanidine hydrochloride was studied under different conditions; a number of kinetics parameters were identified. It was shown high thermal stability, and presence of biocidal properties of the polymers with absence of toxicity. These polymers may be used as biocide substances.

20.1   INTRODUCTION

The creation of modern and safe biocidal polymers is an important direction in modern macromolecular chemistry; however there are a number of specific requirements for structure and molecular weight of new polymers for such use [1–4]. For the most efficient implementation of these requirements, it is necessary to accurately determine and learn synthesis regularities both for initial monomers and polymers based on these monomers. Also an urgent task is to find simpler ways of synthesis the initial monomers, allowing not only reducing the duration and complexity of the process, but also eliminating the use of hazardous substances.

20.2   EXPERIMENTAL PART

All polymers were purified by dialysis. Were used dialysis membranes Spectrapor Membrane MWCO 1000–50,000.

1H NMR spectra were measured with a spectrometer "Bruker MDS-300" (300 MHz) in DMSO-d6 and D_2O at 20°C. Dioxan was used as internal standard in kinetic measurements.

IR spectra of the synthesized monomers and polymers were recorded on a spectrophotometer "Specord M82" Carl Zeiss Jena.

For recording thermograms was used differential scanning calorimeter DTAS-1300 (Russia).

20.3   RESULTS AND DISCUSSION

This chapter presents the results of a synthesis study of new advanced derivatives of acrylic acid – (meth)acryloyl guanidines – containing covalently bonded guanidine groups of different structure, giving opportunity to vary the structure in a wide range, both for the monomers and polymers.

It has been studied methods of (meth)acryloyl guanidine synthesis by reaction guanidine with an acid chloroanhydride, methyl(meth)acrylates and the appropriate acids. It was shown that the most promising for the synthesis of (meth)acryloyl guanidine first two methods are (Scheme 1, R = H or CH$_3$).

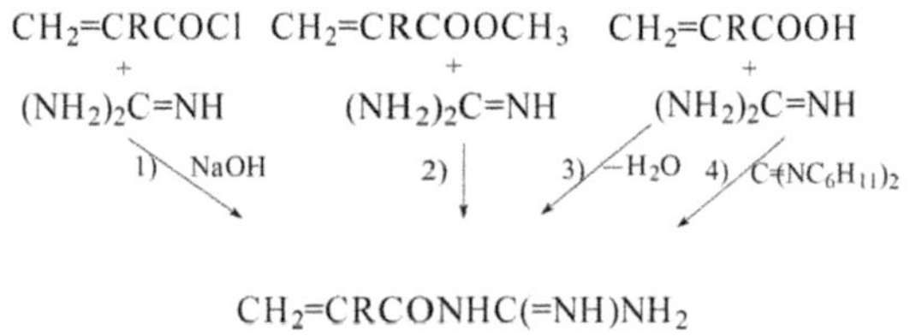

$$CH_2=CRCOCl \quad CH_2=CRCOOCH_3 \quad CH_2=CRCOOH$$
$$+ \qquad\qquad + \qquad\qquad +$$
$$(NH_2)_2C=NH \qquad (NH_2)_2C=NH \qquad (NH_2)_2C=NH$$
$$1)\ NaOH \qquad 2) \qquad 3)\ -H_2O \quad 4)\ C(NC_6H_{11})_2$$

$$CH_2=CRCONHC(=NH)NH_2$$

SCHEME 1

On this basis, it was developed a new method for the synthesis of monomer amide – methacryloyl guanidine (MGU).

As the most convenient method of MGU synthesis, through MMA synthesis (Scheme 2) was selected, which has been investigated under various conditions. It was noted, that the process is complicated by the cyclic products (CP) formation (Scheme 3).

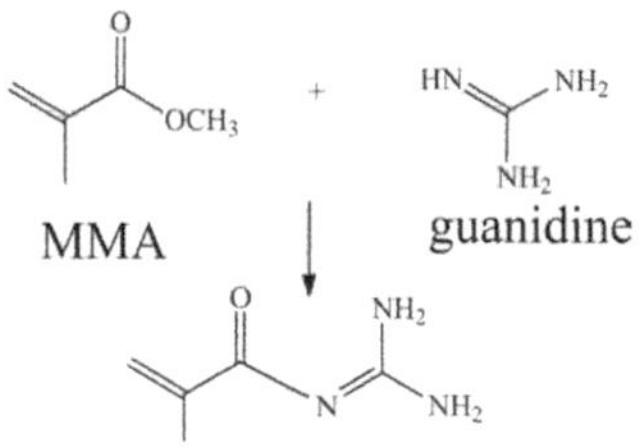

MMA + guanidine

SCHEME 2

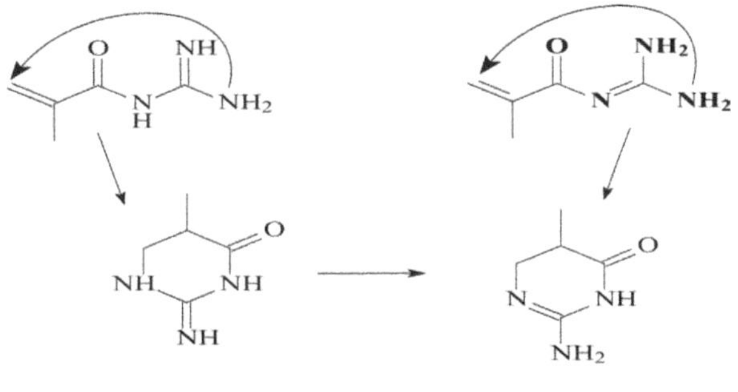

SCHEME 3

TABLE 20.1   Influence of Synthesis Conditions on the Methacryloyl Guanidine Yield

En-try	Synthesis conditions				Yield, mol%	
	Solvent	T,°C	Time, h	C, mol/l	MGU	CP
1	Dioxane	20	63	0.40	49	45
2	Dioxane	20	57	1.18	66	26
3	Dioxane	20	47	0.71	61	34
4	Dioxane	60	3.5	0.73	55	36
5	MMA	20	47	2.31	65	33
6	MMA	40	2	4.59	67	21
7	ACETONE	20	17.5	1.33	64	35
8	ACETONE	20	8	1,4	92	7
9	CH_3CN	20	9	1.38	73	26
10	CH_3CN	60	3.5	1.38	55	45

Analysis of the data presented in Table 20.1, leads to the following conclusions. Best results were obtained when the process was being conducting in acetone (entry 8) and MMA, when it was used both as a solvent and as a reagent (entry 5 and 6). In dioxane MGU formed with sufficiently high yield, but the long stirring and heating of the reaction mixture increase the degree of side processes. Good results were obtained in acetonitrile (entry 9 and 10), but in this case increase of the reaction temperature, which allows to reduce the reaction time, leads to a reduced yield of MGU.

Synthesis of methacryloyl guanidine hydrochloride (MGHC) was carried out by reacting preformed MGU with hydrochloric acid. It was determined that MGU, which had been separated from solvent, subsequently was partially (5–7%) exposed to structural changes, which led to the loss of solubility. In this connection, it is advisable to carry out the synthesis without isolating salts of MGU. It's enough to remove the solid cyclic by-products, which yield is about 10%, by filtration, and then add a hydrochloric acid to obtain a precipitate of MGGH:

$$CH_2{=}\overset{\overset{\displaystyle CH_3}{|}}{C}{-}\overset{\overset{\displaystyle O}{\|}}{C}{-}NH{-}\underset{\underset{\displaystyle \oplus NH_2}{\|}}{C}{-}NH_2 \quad Cl^{\ominus}$$

MAG previously prepared from guanidine hydrochloride (GGH) in a three-step procedure [1, 3, 5]:

a) Sodium ethylate preparation.

$$C_2H_5OH + Na \longrightarrow C_2H_5ONa + 1/2H_2 \uparrow$$

b) Guanidine preparation.

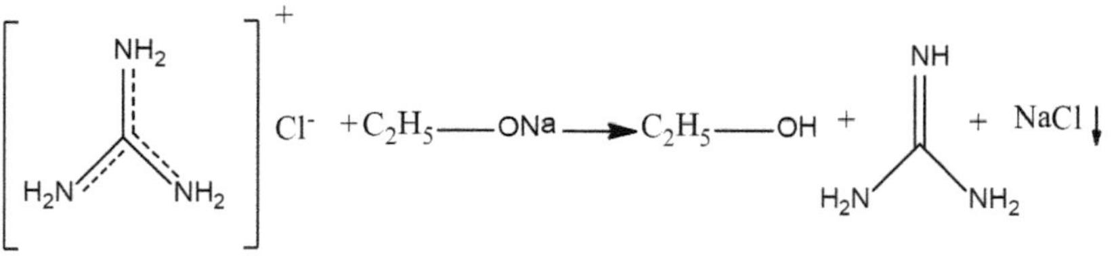

c) Guanidine methacrylate preparation

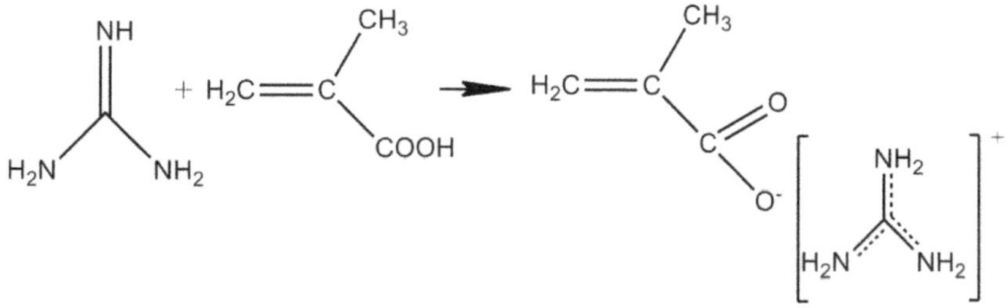

SCHEME 4

This work has been developed and tested with a new method, which allows excluding the stage of guanidine synthesis. The new technique gave the best results (yield 95% vs. 70–75%).

NMR^{1}H and IR spectroscopy data demonstrate that composition and structure of monomer, synthesized in our new method, consistent composition and structure of the monomer obtained by the method of Scheme 4 [6].

In addition, the new technique eliminates the use of ethyl alcohol and excludes work with metallic sodium and gives higher yield of monomer.

Also in this study, we investigated the polymerization of methacrylate guanidine (MAG) and methacryloyl guanidine hydrochloride (MGHC) under various conditions in order to control the structure and other properties of the resulting PMAG and PMGHC. Radical polymerization is carried out in water with ammonium persulfate for 3–9 h at temperatures of

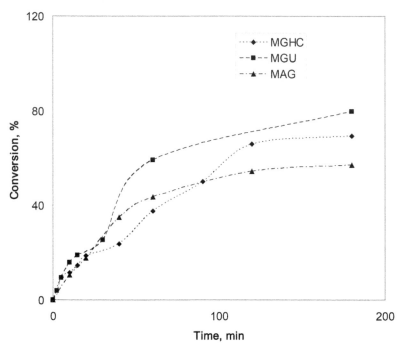

FIGURE 20.1   The dependence of monomer to polymer conversion degree of time.

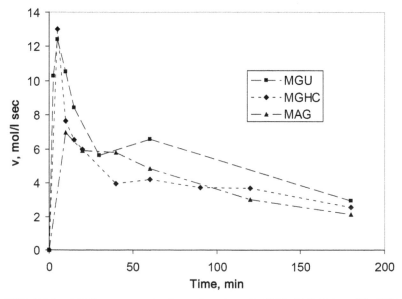

FIGURE 20.2   MAG cure speed of reaction time T=60°C, [M]=0.4 mol/L [PSA] = 5×10^{-3} mol/L.

TABLE 20.2 Kinetic Characteristics of Guanidine Containing Monomers Homopolymerization

№	Monomer	Duration of polymerization 10 min		Duration of polymerization 1 h	
		$k \times 10^4$, c^{-1}	$v \times 10^5$, mol/(L sec)	$k \times 10^4$, sec^{-1}	$v \times 10^5$, mol/(L sec)
1	MGU	2.30	12.73	1.61	6.59
2	MGHC	2.30	10.53	1.15	4.22
3	MAG	1.84	7.00	1.61	4.86
4	MAG*	1.31	6.56	-	-

* determined by dilatometry.

30–80°C, the monomer concentration of 0.3–1.0 mol/L, the concentration of initiator $(2.5–10.0) \times 10^{-3}$ mol/L. Under the same conditions of polymer synthesis conversion is approximately the same for both monomers, whereas the viscosity of the samples PMGHC is higher. It can be explained by the fact that the main chain of PMGHC is positively charged, while the PMAG – negatively. The studies revealed that it is kept main classical patterns: conversion with increasing temperature rises (a range of values 37–91%) and the intrinsic viscosity is reduced (1.50–0.05 dL/g). It was identified a number of kinetic parameters such as rate of polymerization and the rate constant for the initial portion (Figure 20.1 and 20.2, Table 20.2).

According to microbiological tests polymethacrylateguanidine and polymethacryloyl guanidine hydrochloride have sufficiently high bactericidal (relative to strains of *Staphylococcus aureus* and *Escherichia coli*) and low toxicity. DSC showed that the resulting PMAG and PMGHC are thermally stable.

ACKNOWLEDGEMENT

This work was supported by RFBR project number 12-03-00636-a.

KEYWORDS

- methacrylate guanidine
- methacryloyl guanidine hydrochloride
- radical polymerization in water

REFERENCES

1. Sivov, N. A. "Biocide Guanidine containing Polymers: Synthesis, Structure and Properties", Brill Academic Publishers, 2006, p. 151.
2. Sivov, N. A., Martynenko, A. I., Popova, N. I. "Synthesis of new polyfunctional guanidine containing acrylic monomers and polymers." In "Handbook of Condensed Phase Chemistry", Nova Science Publishers, New York. 2011, pp. 307–312.
3. Martynenko, A. I., Khashirova, S. Yu., Malkanduev, Yu. A., Sivov, N. A. "Guanidine containing Polymers: Synthesis, Structure and Properties." Nalchik, M. and V. Kotlyarovs Publishing House, 2008, p. 232 (in Russian).
4. Sivov, N. A. "Nitrogen containing substances. Synthesis, Structure Properties and biological activity." LAP Lambert Acad. Pub. (in Russian), 2014, p. 260.
5. Sivov, N. A., Martynenko, A. I., Kabanova, E. Yu., Popova, N. I., Khashirova S. Yu., Ésmurziev, A. M. Methacrylate and Acrylate Guanidines: Synthesis and Properties. Petroleum Chemistry, Vol. 44, No. 1, 2004, pp. 43–47.
6. Martynenko, A. I., Popova, N. I., Kabanova, E. Yu., Lachinov, M. B., Sivov, N. A. Free-Radical Polymerization of Guanidine Acrylate and Methacrylate and the Conformational Behavior of Growing Chains in Aqueous Solutions." Polymer Science, Ser. A, 2008, Vol. 50, No. 7, pp. 771–780.

CHAPTER 21

NMR ^{1}H AND ^{13}C SPECTRA OF THE 1,1,3-TRIMETHYL-3-(4-METHYL-PHENYL)BUTYL HYDROPEROXIDE IN CHLOROFORM: EXPERIMENTAL VERSUS GIAO CALCULATED DATA

N. A. TUROVSKIJ,[1] E. V. RAKSHA,[1] YU. V. BERESTNEVA,[1] and G. E. ZAIKOV[2]

[1]*Donetsk National University, Universitetskaya Street, 24, Donetsk, 83 001, Ukraine; E-mail: na.turovskij@gmail.com; elenaraksha411@gmail.com*

[2]*Institute of Biochemical Physics, Russian Academy of Sciences, Kosygin Street, 4, Moscow, 117 334, Russian Federation; E-mail: chembio@sky.chph.ras.ru*

CONTENTS

ABSTRACT

NMR ^{1}H and ^{13}C spectra of the 1,1,3-trimethyl-3-(4-methylphenyl)butyl hydroperoxide in chloroform-d have been investigated. Calculation of magnetic shielding tensors and chemical shifts for ^{1}H and ^{13}C nuclei of the 1,1,3-trimethyl-3-(4-methylphenyl)butyl hydroperoxide molecule in the approximation of an isolated particle and considering the solvent influence in the framework of the continuum polarization model (PCM) was carried out. Comparative analysis of experimental and computer NMR spectroscopy results revealed that the GIAO method with B3LYP/6–31G (d, p) level of theory and the PCM approach can be used to estimate the NMR ^{1}H and ^{13}C spectra parameters of the 1,1,3-trimethyl-3-(4-methylphenyl)butyl hydroperoxide.

21.1   INTRODUCTION

Hydroperoxide compounds are widely used as chemical source of the active oxygen species. Variations in their structure allows purposefully create new initiating systems with a predetermined reactivity. Arylalkyl hydroperoxides are useful starting reagents in the synthesis of surface-active peroxide initiators for the preparation of polymeric colloidal systems with improved stability [1]. Thermolysis of arylalkyl hydroperoxides was studied in acetonitrile [2]. NMR ^{1}H spectroscopy has been already used successfully for the experimental evidence of the a complex formation between a 1,1,3-trimethyl-3-(4-methylphenyl)butyl hydroperoxide and tetraalkylammonium bromides in acetonitrile [3–5] and chloroform solution [5].

   Molecular modeling of the peroxide bond homolytic cleavage as well as processes of hydroperoxides association is an additional source of information on the structural effects that accompany these reactions. One of the criteria for the selection of quantum-chemical method for the study of the hydroperoxides reactivity may be reproduction with sufficient accuracy of their NMR ^{1}H and ^{13}C spectra parameters. The aim of this work is a comprehensive study of the 1,1,3-trimethyl-3-(4-methylphenyl)butyl hydroperoxide (ROOH) by experimental ^{1}H and ^{13}C NMR spectroscopy and molecular modeling methods.

21.2 EXPERIMENTAL PART

The 1,1,3-trimethyl-3-(4-methylphenyl)butyl hydroperoxide (ROOH) was purified according to Ref. [1]. Its purity (99%) was controlled by iodometry method. Experimental NMR ^{1}H and ^{13}C spectra of the hydroperoxide solution were obtained by using the Bruker Avance II 400 spectrometer (NMR ^{1}H – 400 MHz, NMR ^{13}C – 100 MHz) at 297 K. Solvent, chloroform-d (CDCl$_3$) was Sigma-Aldrich reagent and was used without additional purification but was stored above molecular sieves before using. Tetramethylsilane (TMS) was internal standard. The hydroperoxide concentration in solution was 0.03 mol·dm^{-3}.

1,1,3-Trimethyl-3-(4-methylphenyl)butyl   hydroperoxide   (4-CH$_3$-C$_6$H$_4$-C(CH$_3$)$_2$-CH$_2$-(CH$_3$)$_2$C-O-OH) NMR ^{1}H (400 MHz, chloroform-d, 297 K, δ ppm, *J*/Hz): 1.00 (s, 6 H, -C(CH$_3$)$_2$OOH), 1.39 (s, 6 H, -C$_6$H$_4$C(CH$_3$)$_2$-), 2.05 (s, 2 H, -CH$_2$-), 2.32 (s, 3 H, CH$_3$-C$_6$H$_4$-), 7.11 (d, *J* = 8.0, 2 H, H-aryl), 7.29 (d, *J* = 8.0, 2 H, H-aryl), 6.77 (s, 1 H, -COOH).

Molecular geometry and electronic structure parameters, as well as harmonic vibrational frequencies of the 1,1,3-trimethyl-3-(4-methylphenyl)butyl hydroperoxide molecule were calculated after full geometry optimization in the framework of B3LYP/6–31G(d, p) and MP2/6–31G(d, p) calculations. The resulting equilibrium molecular geometry was used for total electronic energy calculations by the B3LYP/6–31G(d, p) and MP2/6–31G(d, p) methods. All calculations have been carried out using the Gaussian03 [6] program.

The magnetic shielding tensors (χ, ppm) for ^{1}H and ^{13}C nuclei of the hydroperoxide and the reference molecule were calculated with the MP2/6–31G(d, p) and B3LYP/6–31G(d, p) equilibrium geometries by standard GIAO (Gauge-Independent Atomic Orbital) approach [7]. The calculated magnetic isotropic shielding tensors, χ_i, were transformed to chemical shifts relative to TMS molecule, δ_i, by $\delta_i = \chi_{ref} - \chi_i$, where both, χ_{ref} and χ_i, were taken from calculations at the same computational level. Table 21.1 illustrates obtained χ values for TMS molecule used for the hydroperoxide ^{1}H and ^{13}C nuclei chemical shifts calculations. χ values were also estimated in the framework of 6–311G(d, p) and 6–311++G(d, p) basis sets on the base of MP2/6–31G(d, p) and B3LYP/6–31G(d, p) equilibrium geometries. The solvent effect was considered in the PCM approximation [8, 9]. χ values for magnetically equivalent nuclei were averaged.

TABLE 21.1   GIAO-Magnetic Shielding Tensors for ^1H and ^{13}C Nuclei of the TMS

Nuclei	MP2			B3LYP		
	1	*2*	*3*	*1*	*2*	*3*
The isolated particle approximation						
^1H	31.96	32.08	32.05	31.75	31.96	31.93
^{13}C	207.54	199.71	199.37	191.80	184.13	183.72
Chloroform (PCM approximation)						
^1H	31.95	32.08	32.05	31.75	31.95	31.92
^{13}C	207.86	200.13	199.79	192.08	184.53	184.13

Note: 1 – 6–31G(d, p); 2 – 6–311G(d, p); 3 – 6–311++G(d, p).

Inspecting the overall agreement between experimental and theoretical spectra RMS errors (σ) were used to consider the quality of the ^1H and ^{13}C nuclei chemical shifts calculations. Correlation coefficients (R) were calculated to estimate the agreement between spectral patterns and trends.

21.3   RESULTS AND DISCUSSIONS

21.3.1   EXPERIMENTAL NMR ^1H AND ^{13}C SPECTRA OF THE 1,1,3-TRIMETHYL-3-(4-METHYLPHENYL)BUTYL HYDROPEROXIDE

Experimental NMR ^1H and ^{13}C spectra of the 1,1,3-trimethyl-3-(4-methylphenyl)butyl hydroperoxide were obtained from chloroform-d solution. The concentration of the hydroperoxide in sample was 0.03 mol·dm^{-3}. Parameters of the experimental NMR ^1H and ^{13}C spectra of the ROOH are listed in Tables 21.2 and 21.3 correspondingly.

Comparing obtained results with those in acetonitrile-d$_3$ [10], it should be noted that signals shift toward the strong field is observed in

TABLE 21.2   Experimental Chemical Shifts of the 1,1,3-Trimethyl-3-(4-Methylphenyl)
Butyl Hydroperoxide NMR ^{1}H Spectra in Chloroform-d

Proton group		δ, ppm
H1	$-CH_2-$	2.05
H2	$-C(CH_3)_2OOH$	1.00
H3	$-C_6H_4C(CH_3)_2-$	1.39
H4	H-aryl	7.11
H5		7.29
H6	$CH_3-C_6H_4-$	2.32
H7	$-CO-OH$	6.77

TABLE 21.3   Experimental Chemical Shifts of the 1,1,3-Trimethyl-3-(4-Methylphenyl)
Butyl Hydroperoxide NMR ^{13}C Spectra in Chloroform-d

Carbon group		δ, ppm
C1	$-CO-OH$	83.93
C2	$-CH_2-$	50.71
C3	$-C(CH_3)_2OOH$	25.98
C4	$-C_6H_4C(CH_3)_2-$	37.03
C5	$-C_6H_4C(CH_3)_2-$	30.91
C6		146.55
C7	C-aryl	125.81
C8		128.81
C9		135.01
C10	$CH_3-C_6H_4-$	20.86

the spectrum of the hydroperoxide with increasing solvent polarity. On the other hand, the hydroperoxide moiety proton signal is observed in the weak field with the solvent polarity increasing. The -CO-OH group proton appears at 6.77 ppm in chloroform-d, and in the more polar acetonitrile-d$_3$ it was found at 8.51 ppm.

Ten signals for the hydroperoxide carbon atoms are observed in the ROOH ^{13}C NMR spectrum (Table 21.3). Signal of the carbon atom bonded with a hydroperoxide group shifts slightly to the stronger fields with the solvent polarity increasing, while the remaining signals are shifted to weak fields.

A linear dependence between the ^{1}H and ^{13}C chemical shifts values of the hydroperoxide (except for the -CO-OH group proton) is observed in the studied solvents (Figure 21.1). This is consistent with authors [11], who showed linear correlation between the chemical shifts values in chloroform-d and dimethylsulphoxide-d$_6$ for a large number of organic compounds of different classes.

21.3.2 MOLECULAR MODELING OF THE 1,1,3-TRIMETHYL-3-(4-METHYLPHENYL)BUTYL HYDROPEROXIDE STRUCTURE AND NMR 1H AND 13C SPECTRA BY MP2 AND B3LYP METHODS

The hydroperoxide molecule geometry optimization in the framework of MP2/6–31G(d, p) and B3LYP/6–31G(d, p) methods was carried out as the first step of the hydroperoxide NMR ^{1}H and ^{13}C spectra modeling. Initial hydroperoxide configuration chosen for calculations was those one obtained by semiempirical AM1 method and used recently for the hydroperoxide O-O bond homolysis [2] as well as complexation with Et$_4$NBr [4, 12] modeling. The main parameters of the hydroperoxide fragment molecular geometry obtained in the isolated particle approximation within the framework of MP2/6–31G(d, p) and B3LYP/6–31G(d, p) levels of theory are presented in Table 21.4. Peroxide bond O-O is a reaction center in this type of chemical initiators thus the main attention was focused on

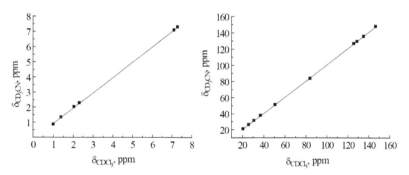

FIGURE 21.1 Acetonitrile-d$_3$ versus chloroform-d experimental ^{1}H (a) and ^{13}C (b) chemical shifts (relative to TMS) of the 1,1,3-trimethyl-3-(4-methylphenyl)butyl hydroperoxide (except for the -CO-OH group proton).

TABLE 21.4   Molecular Geometry Parameters of the 1,1,3-Trimethyl-3-(4-Methylphenyl)Butyl Hydroperoxide -CO-OH Moiety

Parameter	MP2/6–31G(d, p)	B3LYP/6–31G(d, p)	Experiment*
l_{O-O}, Å	1.473	1.456	1.473
l_{C-O}, Å	1.459	1.465	1.443
l_{O-H}, Å	0.970	0.971	0.990
C-O-O, °	108.6	110.0	109.6
O-O-H, °	98.2	99.9	100.0
C-O-O-H, °	112.4	109.1	114.0

*Note: experimental values are those for *tert*-butyl hydroperoxide [13].

the geometry of -CO-OH fragment. The calculation results were compared with known experimental values for the *tert*-butyl hydroperoxide [13], and appropriate agreement between calculated and experimental parameters can be seen in the case of MP2/6–31G(d, p) method.

Calculation of ¹H and ¹³C chemical shifts of the hydroperoxide was carried out by GIAO method in the approximation of an isolated particle as well as in chloroform within the PCM model, which takes into account the nonspecific solvation. Equilibrium hydroperoxide geometries obtained in the framework of MP2/6–31G(d, p) and B3LYP/6–31G(d, p) levels of theory for the isolated particle approximation were used in all cases (Figure 21.2).

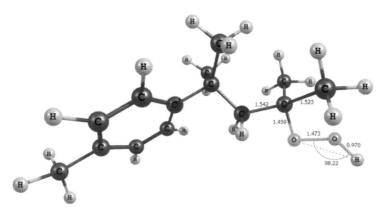

FIGURE 21.2   The 1,1,3-trimethyl-3-(4-methylphenyl)butyl hydroperoxide structural model with corresponding atom numbering (MP2/6–31G(d, p) method).

The chemical shift values (δ, ppm) for ^{1}H and ^{13}C nuclei in the hydroperoxide molecule were evaluated on the base of calculated magnetic shielding constants (χ, ppm). TMS was used as standard, for which the molecular geometry optimization and χ calculation were performed using the same level of theory and basis set. Values of the ^{1}H and ^{13}C chemical shifts were found as the difference of the magnetic shielding tensors of the corresponding TMS and hydroperoxide nuclei (Tables 21.5 and 21.6).

Concerning the spectral pattern of protons, inspection of Table 21.5 data reveals that the patterns of ^{1}H spectra of the hydroperoxide are correctly reproduced at all computational levels used in the study. For

TABLE 21.5 ^{1}H NMR GIAO Chemical Shifts (δ, ppm) of the 1,1,3-Trimethyl-3-(4-Methylphenyl) Butyl Hydroperoxide

Nuclei	MP2			B3LYP			4
	1	2	3	1	2	3	
The isolated particle approximation							
H1	1.60	1.63	1.65	1.57	1.56	1.50	2.05
H2	1.44	1.44	1.47	1.40	1.40	1.42	1.00
H3	1.46	1.47	1.50	1.40	1.38	1.41	1.39
H4	7.54	7.58	7.64	7.32	7.33	7.34	7.11
H5	7.42	7.57	7.63	7.16	7.30	7.37	7.29
H6	2.25	2.39	2.45	2.18	2.33	2.36	2.32
H7	6.68	6.56	6.77	5.89	5.76	5.90	6.77
σ	0.09	0.10	0.12	0.18	0.21	0.19	-
R	0.995	0.995	0.995	0.989	0.986	0.988	-
Chloroform (PCM approximation)							
H1	1.56	1.60	1.62	1.53	1.52	1.47	2.05
H2	1.47	1.47	1.51	1.43	1.43	1.46	1.00
H3	1.49	1.50	1.53	1.42	1.40	1.43	1.39
H4	7.62	7.67	7.74	7.39	7.42	7.43	7.11
H5	7.50	7.67	7.73	7.23	7.38	7.47	7.29
H6	2.29	2.43	2.49	2.22	2.37	2.40	2.32
H7	7.14	7.03	7.24	6.35	6.22	6.37	6.77
σ	0.13	0.14	0.19	0.10	0.12	0.12	-
R	0.995	0.995	0.995	0.993	0.995	0.984	-

Note: *1* – 6-31G(d, p); *2* – 6-311G(d, p); *3* – 6-311++G(d, p); *4* – experiment in chloroform-d.

TABLE 21.6 ^{13}C NMR GIAO Chemical Shifts (δ, ppm) of the 1,1,3-Trimethyl-3-(4-Methylphenyl)Butyl Hydroperoxide

Nuclei	MP2			B3LYP			4
	1	2	3	1	2	3	
The isolated particle approximation							
C1	83.61	86.87	88.24	85.77	90.90	92.04	83.93
C2	53.50	57.48	57.42	53.23	57.70	57.00	50.71
C3	26.46	26.71	26.62	24.62	25.27	24.93	25.98
C4	37.04	40.23	40.37	41.14	44.66	44.49	37.03
C5	30.10	30.93	31.03	28.57	29.78	29.75	30.91
C6	141.39	153.47	153.93	144.46	158.38	158.37	146.55
C7	116.91	125.90	126.29	119.80	130.58	131.03	125.81
C8	127.07	137.37	137.97	130.75	142.40	143.32	128.81
C9	121.55	130.79	131.41	123.68	134.44	135.07	135.01
C10	22.98	23.82	23.75	21.84	23.25	22.84	20.86
σ	*23.49*	*13.13*	*15.39*	*11.99*	*41.22*	*44.24*	-
R	*0.997*	*0.997*	*0.997*	*0.996*	*0.996*	*0.996*	-
Chloroform (PCM approximation)							
C1	84.29	87.74	89.27	86.44	91.86	93.14	83.93
C2	53.69	57.73	57.64	53.37	57.89	57.17	50.71
C3	26.62	26.99	26.89	24.73	25.52	25.17	25.98
C4	37.46	40.74	40.92	41.54	45.20	45.07	37.03
C5	30.18	31.10	31.20	28.59	29.90	29.88	30.91
C6	142.14	154.40	154.90	145.12	159.23	159.25	146.55
C7	117.33	126.51	126.91	120.11	131.14	131.60	125.81
C8	127.92	138.43	139.03	131.56	143.43	144.34	128.81
C9	121.90	131.35	131.91	123.85	134.81	135.37	135.01
C10	23.01	23.96	23.88	21.81	23.36	22.93	20.86
σ	*27.86*	*25.63*	*28.66*	*20.82*	*59.16*	*63.32*	-
R	*0.997*	*0.997*	*0.997*	*0.996*	*0.996*	*0.996*	-

Note: 1 – 6-31G(d, p); 2 – 6-311G(d, p); 3 – 6-311++G(d, p); 4 – experiment in chloroform-d.

MP2 method the expansion of the basis set leads to slightly worse repro-
ducing of protons chemical shifts values (except for hydroperoxide group
proton) in the case of an isolated particle approximation. But as for

CO-OH group proton the best reproduction of the experimental δ value is observed when 6–311++G(d, p) basis set is used (6.77 ppm). For results obtained at other computational level, we note that with respect to spectral patterns and trends, results obtained with the computationally less expensive B3LYP optimized geometry are very similar to those obtained with the MP2 calculations. When passing to the calculations in the PCM mode solvation accounting leads to more correct results for the MP2 and B3LYP methods. The lowest σ value is obtained for 6–31G(d, p) basis set.

There is a linear correlation between the experimental and calculated with solvent influence accounting δ values for the hydroperoxide ^1H nuclei in chloroform (see Figure 21.3). The correlation coefficients (R) corresponding to obtained dependences are shown in Table 21.5. The best R-values for MP2 and B3LYP methods are observed for 6-311G(d, p) basis set and further basis set extension leads to slightly worse values. Joint consideration of σ and R-values indicates B3LYP/6-311G(d, p) method is the best among all used combinations. Nevertheless is should be noted that the B3LYP with 6-31G(d, p) basis set yields the similar results.

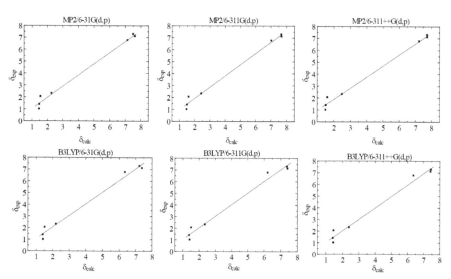

FIGURE 21.3 Experimental (δ_{exp}) versus GIAO calculated (δ_{calc}) ^1H chemical shifts (relative to TMS) of the 1,1,3-trimethyl-3-(4-methylphenyl)butyl hydroperoxide in chloroform.

The correct spectral pattern for the hydroperoxide NMR ¹³C spectrum was obtained for all methods and basis sets used within the isolated molecule approximation as well as solvation accounting (see Table 21.6). Exceptions are aromatic hydrocarbons C8 and C9, which signals are interchanged for all calculations.

The best-reproduced experimental chemical shift value for the carbon atom of the CO-OH group (83.93 ppm) is observed in the case of MP2/6–31G(d, p) (83.61 and 84.29 ppm), the B3LYP with the same basis set gives slightly worse value (85.77 and 86.44 ppm). Basis set extension to 6–311++G(d, p) leads to a deterioration of the calculation results.

Linear relationships between the experimental parameters of the NMR ¹³C spectrum and the calculated values δ_{calc} for the hydroperoxide ¹³C nuclei (see Figure 21.4) have been obtained for both methods and all basis sets. Sufficiently high values of correlation coefficients (see Table 6) correspond to these dependences. Joint account of σ and R values indicates possibility of B3LYP method with 6–31G(d, p) basis set using for the calculation of the hydroperoxide ¹³C nuclei chemical shifts.

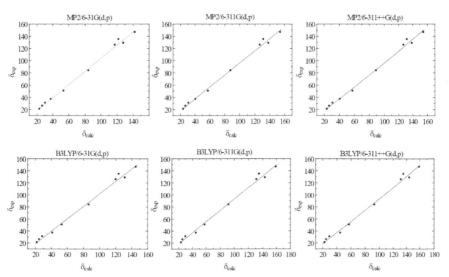

FIGURE 21.4 Experimental (δ_{exp}) versus GIAO calculated (δ_{calc}) ¹³C chemical shifts (relative to TMS) of the 1,1,3-trimethyl-3-(4-methylphenyl)butyl hydroperoxide in chloroform.

21.4   CONCLUSIONS

A comprehensive study of the 1,1,3-trimethyl-3-(4-methyl-phenyl)butyl hydroperoxide by experimental NMR ^{1}H and ^{13}C spectroscopy and molecular modeling methods was performed. A comparative assessment of the ^{1}H and ^{13}C nuclei chemical shifts calculated by GIAO in various approximations. For NMR ^{1}H and ^{13}C spectra of the hydroperoxide in chloroform MP2 and B3LYP methods approximations with 6-31G(d, p), 6-311G(d, p), and 6-311++G(d, p) basis sets allow to obtain the correct spectral pattern. A linear correlation between the calculated and experimental values of the ^{1}H and ^{13}C chemical shifts for the studied hydroperoxide molecule. In both cases, the B3LYP combined with 6-31G(d, p) basis set allows to get a better agreement between the calculated and experimental data as compared to the MP2 results.

KEYWORDS

- **NMR spectroscopy**
- **chemical shift**
- **magnetic shielding constant**
- **GIAO**
- **molecular modeling**
- **1,1,3-trimethyl-3-(4-methylphenyl)butyl hydroperoxide**

REFERENCES

1. Kinash, N.I., Vostres, V. B. Sinthesys of the γ-aryl containing peroxides 2-methyl-4-pentane derivatives. Visnyk Natsional'nogo Universitetu "Lvivska Polytechnika", 2003, No 529, 124–128.
2. Turovsky, M. A., Raksha, O. V., Opeida, I. O., Turovska, O. M., Zaikov, G. E. Molecular modeling of aralkyl hydroperoxides homolysis. Oxidation Communications. 2007, Vol. 30, No 3, 504–512.
3. Turovskij, N. A., Raksha, E. V., Berestneva Yu.V., Zubritskij, M.Yu. Formation of 1,1,3-trimethyl-3-(4-methylphenyl)butyl hydroperoxide complex with tetrabutylammonium bromide. Russian, J. Gen. Chem., 2014, Vol. 84, No. 1, pp. 16–17.
4. Berestneva Yu.V., Raksha, E. V., Turovskij, N.A., Zubritskij, M.Yu. Interaction of the 1,1,3-trimethyl-3-(4-methylphenyl)butyl hydroperoxide with tetraethylammonium bromide. Actual problems of magnetic resonance and its application: program lecture

notes proceedings of the XVII International Youth Scientific School (Kazan, 22–27 June 2014). edited by, M. S. Tagirov (Kazan Federal University), V. A. Zhikharev (Kazan State Technological University). Kazan: Kazan University, 2014, 165 p., 113–116.

5. Turovskij, N.A., Berestneva Yu.V., Raksha E.V., Opeida, J.A, Zubritskij, M.Yu. Complex formation between hydroperoxides and Alk$_4$NBr on the base of NMR spectroscopy investigations. Russian Chemical Bulletin, 2014, No 8, 1717–1721.

6. Gaussian 03, Revision, B.01, Frisch, M. J., Trucks, G. W., Schlegel, H. B., Scuseria, G. E., Robb, M. A., Cheeseman, J. R., Montgomery, J. A., Jr., Vreven, T., Kudin, K. N., Burant, J. C., Millam, J. M., Iyengar, S. S., Tomasi, J., Barone, V., Mennucci, B., Cossi, M., Scalmani, G., Rega, N., Petersson, G. A., Nakatsuji, H., Hada, M., Ehara, M., Toyota, K., Fukuda, R., Hasegawa, J., Ishida, M., Nakajima, T., Honda, Y., Kitao, O., Nakai, H., Klene, M., Li, X., Knox, J. E., Hratchian, H. P., Cross, J. B., Adamo, C., Jaramillo, J., Gomperts, R., Stratmann, R. E., Yazyev, O., Austin, A. J., Cammi, R., Pomelli, C., Ochterski, J. W., Ayala, P. Y., Morokuma, K., Voth, G. A., Salvador, P., Dannenberg, J. J., Zakrzewski, V. G., Dapprich, S., Daniels, A. D., Strain, M. C., Farkas, O., Malick, D. K., Rabuck, A. D., Raghavachari, K., Foresman, J. B., Ortiz, J. V., Cui, Q., Baboul, A. G., Clifford, S., Cioslowski, J., Stefanov, B. B., Liu, G., Liashenko, A., Piskorz, P., Komaromi, I., Martin, R. L., Fox, D. J., Keith, T., Al-Laham, M. A., Peng, C. Y., Nanayakkara, A., Challacombe, M., Gill, P. M. W., Johnson, B., Chen, W., Wong, M. W., Gonzalez, C., Pople, J. A., Gaussian, Inc., Pittsburgh PA, 2003.

7. Wolinski, K., Hinton, J. F., Pulay, P. Efficient implementation of the gauge-independent atomic orbital method for NMR chemical shift calculations. J. Am. Chem. Soc. 1990, Vol. 112(23), 8251–8260.

8. Mennucci, B., Tomasi, J. Continuum solvation models: A new approach to the problem of solute's charge distribution and cavity boundaries. J. Chem. Phys. 1997, Vol. 106, 5151–5158.

9. Cossi, M., Scalmani, G., Rega, N., Barone, V. New developments in the polarizable continuum model for quantum mechanical and classical calculations on molecules in solution. J. Chem. Phys. 2002, Vol. 117, 43–54.

10. Raksha, E. V., Berestneva Yu.V., Turovskij, N.A., Zubritskij M.Yu. Quantum chemical modeling of the 1,1,3-trimethyl-3-(4-methyl-phenyl)butyl hydroperoxide NMR ^{1}H and ^{13}C spectra. Naukovi pratsi DonNTU. Ser.: Khimiya i khimichna tehnologiya. 2014, Iss. 1(22), 150–156.

11. Abraham, R. J., Byrne, J. J., Griffiths, L., Perez, M. ^{1}H chemical shifts in NMR: Part 23, the effect of dimethyl sulphoxide versus chloroform solvent on ^{1}H chemical shifts. Magn. Reson. Chem. 2006, Vol. 44, 491–509.

12. Turovskij, N. A., Raksha, E. V., Berestneva Yu.V., Pasternak, E. N., Zubritskij, M. Yu., Opeida, I. A., Zaikov, G. E. Supramolecular Decomposition of the Aralkyl Hydroperoxides in the Presence of Et$_4$NBr. in: Polymer Products and Chemical Processes: Techniques, Analysis Applications, Eds. R. A. Pethrick, E. M. Pearce, G. E. Zaikov, Apple Academic Press, Inc., Toronto, New Jersey, 2013, 270.

13. Kosnikov, A. Yu., Antonovskii, V. L., Lindeman, S. V., Antipin, M. Yu., Struchkov Yu. T., Turovskii, N. A., Zyat'kov, I. P. X-ray crystallographic and quantum-chemical investigation of tert-butyl hydroperoxide. Theoretical and Experimental Chemistry. 1989, Vol. 25. Iss. 1, 73–77.

PART IV

SPECIAL TOPICS

CHAPTER 22

RHEOLOGICAL PROPERTIES OF THE IRRIGATION LIQUID IN THE CLEANING PROCESS OF GAS EMISSIONS

R. R. USMANOVA[1] and G. E. ZAIKOV[2]

[1]Ufa State Technical University of Aviation, 12 Karl Marks Str., Ufa 450100, Bashkortostan, Russia; E-mail: Usmanovarr@mail.ru

[2]N.M. Emanuel Institute of Biochemical Physics, Russian Academy of Sciences, 4 Kosygin Str., Moscow 119334, Russia; E-mail: chembio@sky.chph.ras.ru

CONTENTS

ABSTRACT

The study of fluid flow and particle separation in rotoklon allowed to consider in detail all the stages of the process of hydrodynamic interaction of phases in devices shock-inertial action. Define the boundary degree of recirculation irrigation liquid, which ensures stable operation rotoklon. It is established that the reduction of fine particle separation efficiency occurs with a decrease in viscosity irrigation liquid.

22.1   INTRODUCTION

One of the problems of gas cleaning devices is to provide apparatus of intense action with high capacity for gas. This is associated with a reduction in the dimensions of the gas cleaning systems.

In these conditions, due to the high relative velocity of the liquid and gas phases, a decisive influence on the effect of dust collection have mechanisms: inertial and direct capture of particles. This process is implemented in a shock-inertial dust collector, which include investigated apparatus.

In the literature practically there are no data on effect of viscosity of a trapping liquid on dust separation process. Therefore, one of the purposes of our work was revealing of effect of viscosity of a liquid on efficiency of a dust separation.

22.2   THE URGENCY OF THE PROBLEM

The problem formulation leans on following rules. In the conditions of full circulation of a liquid, at constant geometrical sizes of a deduster it is possible to secure with a constancy of operational parameters is a relative speed of traffic of a liquid and an aerosol, concentration of a dust in gas, a superficial tension of a liquid or an angle of wetting of a dust. Concentration of a dust growing in a time in a liquid conducts to unique essential change – to increase in its viscosity. After excess of certain concentration suspension loses properties of a Newtonian fluid. Deduster working conditions at full circulation of a liquid are approached to what can be gained in the periodical regime when at maintenance fresh water is not inducted into dust-collecting plant. Collected in the apparatus, the dust detained by a liquid, compensates volume losses of the liquid necessary on moistening of passing gas and its ablation. In the literature there are no works, theoretically both experimentally presenting effect of viscosity and effect of rheological properties of slurry on efficiency of a dust separation. As it seems to us, a motive is that fact that in the capacity of operating fluid water is usually used, and dedusters work, predominantly, at constant temperatures. Simultaneously, at use of partial circulation certain level of concentration of a dust in a liquid is secured. In turn, accessible dependences in the literature specify in insignificant growth of viscosity of slurry even at raise of its concentration for some percent.

The reasoning's proving possibility of effect of viscosity of slurry on efficiency of a dust separation, it is possible to refer to as on the analysis of the basic mechanisms influencing sedimentation of corpuscles on an interphase surface, and on conditions of formation of this developed surface of a liquid. Transition of corpuscles of a dust from gas in a liquid occurs, mainly, as a result of the inertia affecting, effect of "sticking" and diffusion. Depending on type of the wet-type collector of a corpus of a dust deposit on a surface of a liquid which can be realized in the form of drops, moving in a stream of an aerosol, the films of a liquid generated in the apparatus, a surface of the gas vials formed in the conditions of a barbotage and moistened surfaces of walls of the apparatus.

In the monography [3] effect of various mechanisms on efficiency of sedimentation of corpuscles of a dust on a liquid surface is widely presented. The description of mechanisms and their effect on efficiency of a dust separation can be found practically in all monographies, for example, [4, 13] concerning a problem wet clearings of gas emissions of gasses. In the literature of less attention it is given questions of formation of surfaces of liquids and their effect on efficiency of a dust separation.

Observing the mechanism of the inertia act irrespective of a surface of the liquid entraining a dust, predominantly it is considered that for hydrophilic types of a dust collision of a part of a solid with a liquid surface to its equivalent immediate sorption by a liquid, and then immediate clearing and restoration of the surface of a liquid for following collisions. In case of a dust badly moistened, the time necessary for sorbtion of a corpuscle by a liquid, can be longer, than a time after which the corpuscle will approach to its surface. Obviously, it is at the bottom of decrease in possibility of a retardation of a dust by a liquid because of a recoil of the corpuscle going to a surface, from a corpuscle, which is on it. It is possible to consider this effect real as in the conditions of a wet dust separation with a surface of each fluid element impinges more dust, than it would be enough for monolayer formation. Speed of sorbtion of corpuscles of a dust can be a limiting stage of a dust separation.

Speed of sorbtion of a corpuscle influences not only its energy necessary for overcoming of a surface tension force, but also and its traverse speed in the liquid medium, depending on its viscosity and rheological properties. Efficiency of dust separation Kabsch [5] connects with speed of ablation of a dust a liquid, having presented it as weight ms, penetrating in unit of time through unit of a surface and in depth of a liquid as a result of a collision of grains of a dust with this surface:

$$r = \frac{m_s}{A \cdot t}$$

Giving to shovels of an impeller sinusoidal a profile allows to eliminate breakaway a stream breakaway on edges. Thus there is a flow of an entrance section of a profile of blades with the big constant speed and increase in ricochets from a shaped part of blades in terms of which it is possible to predict insignificant increase in efficiency of clearing of gas.

Speed of linkage of a dust a liquid depends on physicochemical properties of a dust and its ability to wetting, physical and chemical properties of gas and operating fluid, and also concentration of an aerosol. Wishing to confirm a pushed hypothesis, Kabsch [5] conducted the research concerning effect of concentration speed of linkage of a dust by a liquid. The increase in concentration of a dust in gas called some increase in speed of linkage, however to a lesser degree, than it follows from linear dependence.

The cores for technics of a wet dust separation of model Semrau, Barth'a and Calvert'a do not consider effect of viscosity of slurry on effect of a dust separation. In-process Pemberton'a [6] it is installed that in case of sedimentation of the corpuscles, which are not moistened on drops, their sorbtion in a liquid is obligatory, and their motion in a liquid submits to principle Stokes'a.

The traverse speed can characterize coefficient of resistance to corpuscle motion in a liquid, so, and a dynamic coefficient of viscosity of a liquid. Possibility of effect of viscosity of a liquid on efficiency of capture of corpuscles of a dust a drop by simultaneous Act of three mechanisms: the inertia, "capture" the semiempirical equation Slinna [6] considering the relation of viscosity of a liquid to viscosity of gas also presents.

In general, it is considered that there is a certain size of a drop [14] at which optimum conditions of sedimentation of corpuscles of a certain size are attained, and efficiency of subsidence of corpuscles of a dust on a drop sweepingly decreases with decrease of a size of these corpuscles.

Jarzkbski and Giowiak [9], analyzing work of an impact-sluggish deduster have installed that in the course of a dust separation defining role is played by the phenomenon of the inertia collision of a dust with water drops. Efficiency of allocation of corpuscles of a dust decreases together with growth of sizes of the drops oscillated in the settling space, in case of a generating of drops compressed air, their magnitude is defined by equation Nukijama and Tanasawa [10] from which follows that drops to those more than above value of viscosity of a liquid phase. Therefore, viscosity growth can call reduction of efficiency of a dust separation.

The altitude of a layer of the dynamic foam formed in dust-collecting plant at a certain relative difference of speeds of gas and liquid phases, decreases in process of growth of viscosity of a liquid [15] that calls decrease in efficiency of a dust separation, it is necessary to consider that the similar effect refers to also to a layer of an intensive barbotage and the drop layer partially strained in dust removal systems.

Summarizing it is possible to assert that in the literature practically there are no data on effect of viscosity of a trapping liquid on dust separation process. Therefore, one of the purposes of our work was extraction of effect of viscosity of a liquid on efficiency of a dust separation.

22.2.1   THE PURPOSE AND OBJECTIVES OF RESEARCH

The conducted research had a main objective acknowledging of a hypothesis on existence of such boundary concentration of slurry at which excess the overall performance of the dust removal apparatus decreases.

The concept is devised and the installation, which is giving the chance to implementation of planned research is mounted. Installation had systems of measurement of the general and fractional efficiency and typical systems for measurement of volume flow rates of passing gas and water resistances. The device of an exact proportioning of a dust, and also the air classifier separating coarse fractions of a dust on an entry in installation is mounted. Gauging of measuring systems has secured with respective repeatability of the gained results.

22.3   LABORATORY FACILITY AND TECHNIQUE OF CONDUCTING OF EXPERIMENT

Laboratory facility basic element is the deduster of impact-sluggish act – a rotoklon c adjustable guide vanes [11] (*see*, Figure 22.1). An aerosol gained by dust introduction in the pipeline by means of the batcher. Application of the batcher with changing productivity has given the chance to gain the set concentration of a dust on an entry in the apparatus.

Have been investigated a dust, discriminated with the wettability (a talcum powder the ground, median diameter is equal $\delta_{50} = 25$ microns white black about $\delta_{50} = 15$ microns solubility in water of $10^{-3}\%$ on weight ($25°C$) and a chalk powder).

The gas-dispersed stream passed shovels of an impeller 7 in a working zone of the apparatus, whence through the drip pan 8 cleared, was inferred outside. Gas was carried by means of the vacuum pump 10, and its charge measured by means of a diaphragm 1. A gas rate, passing through

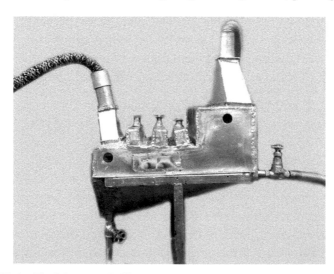

FIGURE 22.1   The laboratory facility.

installation, controlled, changing quantity of air sucked in the pipeline before installation. The composition of each system includes group of three sondes mounted on vertical sections of pipelines, on distance about 10 diameters from the proximal element changing the charge. The taken test of gas went on the measuring fine gage strainer on which all dust containing in test separated. For this purpose used fine gage strainer. In the accepted solution have applied system of three measuring sondes, which have been had in pipelines so that in the minimum extent to change a regime of passage of gas and to select quantity of a dust necessary for the analysis. The angle between directions of deducing of sondes made 120°, and their ends placed on such radiuses that surfaces of rings from which through a sonde gas was sucked in, were in one plane. It has allowed to scale down a time of selection of test and gave average concentration of a dust in gas pipeline cross-section.

Fractional composition of a dust on an entry and an exit from the apparatus measured by means of analogous measuring systems, chapter 10.

For definition of structurally mechanical properties of slurry viscosity RV-8 (Figure 22.2 see) has been used. The viscosity gage consists of the internal twirled cylinder (rotor) (r = 1.6 centimeter) and the external motionless cylinder (stator) (r = 1.9 centimeter), having among themselves a positive allowance of the ring form with a size 0.3 see the Rotor

FIGURE 22.2   Measurement of viscosity of slurry.

is resulted in twirl by means of the system consisting of the shaft, a pulley (T_o = 2.23 centimeter), filaments, blocks and a cargo. To the twirl termination apply a brake. The twirled cylinder has on a division surface on which control depth of its plunging in slurry.

The gained slurry in number of 30 sm³ (in this case the rotor diving depth in sludge makes 7 sm) fill in in is carefully the washed out and dry external glass which put in into a slot of a cover and strengthen its turn from left to right. After that again remove the loaded cylinder that on a scale of the internal cylinder precisely to define depth of its plunging in sludge. Again, fix a glass and on both cups put the minimum equal cargo (on 1), fix the spigot of a pulley by means of a brake and reel up a filament, twirling a pulley clockwise. It is necessary to watch that convolutions laid down whenever possible in parallel each other.

Install an arrow near to any division into the limb and, having hauled down a brake, result the internal cylinder in twirl, fixing a time during which the cylinder will make 4–6 turns. After the termination of measurements fix a brake and reel up a filament. Measurement at each loading spends not less than three times. Experiences repeat at gradual increase in a cargo on 2 gr. until it is possible to fix a time of an integer of turns precisely enough. After the termination of measurements remove a glass, drain from it sludge, wash out water, from a rotor sludge drain a wet rag then both cylinders are dry wiped and leave the device in the collected aspect.

After averaging of the gained data and calculation of angular speed the schedule of dependence of speed of twirl from the enclosed loading is under construction.

Viscosity is defined by formula [15]:

$$\eta = \frac{(R_2^2 - R_1^2)Gt}{8\pi^2 LR_1^2 R_2^2 L} \tag{1}$$

or

$$\eta = \frac{kGt}{L} \tag{2}$$

22.4 DISCUSSION OF RESULTS OF RESEARCH

For each dust used in research dependence of general efficiency of a dust separation on concentration of slurry and the generalizing schedule of dependence of fractional efficiency on a corpuscle size is presented. Other schedule grows out of addition fractional efficiency of a dust separation for various, presented on the schedule, concentration of slurry. In each case the first measurement of fractional efficiency is executed in the beginning of the first measuring series, at almost pure water in a deduster.

Analyzing the gained results of research of general efficiency of a dust separation, it is necessary to underline that in a starting phase of work of a rotoklon at insignificant concentration of slurry for all used in dust research components from 93.2% for black to 99.8% for a talc powder are gained high efficiency of a dust separation. Difference of general efficiency of trapping of various types of a dust originates because of their various fractional composition on an entry in the apparatus, and also because of the various form of corpuscles, their dynamic wettability and density. The gained high values of general efficiency of a dust separation testify to correct selection of constructional and operational parameters of the studied apparatus and specify in its suitability for use in technics of a wet dust separation.

The momentous summary of the spent research was definition of boundary concentration of slurry various a dust after which excess general efficiency of a dust separation decreases. Value of magnitude of boundary concentration, as it is known, is necessary for definition of the maximum extent of recirculation of an irrigating liquid. As appears from presented in Figure 22.3–22.6 schedules, dependence of general efficiency of a dust separation on concentration of slurry, accordingly, for a powder of talc, a chalk and white black is available possibility of definition of such concentration.

Boundary concentration for a talcum powder – 36%, white black – 7%, a chalk – of 18% answer, predominantly, to concentration at which slurries lose properties of a Newtonian fluid.

The conducted research give the grounds to draw deductions that in installations of impact-sluggish type where the inertia mechanism is the

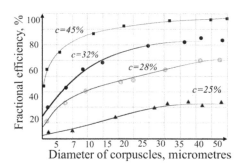

FIGURE 22.3 Dependence of fractional efficiency on diameter of corpuscles of a talcum powder and their concentration in a liquid.

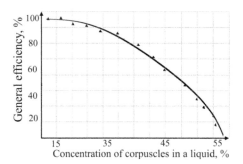

FIGURE 22.4 Dependence of general efficiency of concentration in a liquid of corpuscles of a talcum powder.

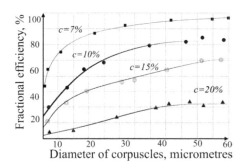

FIGURE 22.5 Dependence of fractional efficiency on diameter of corpuscles of white black and their concentration in a liquid.

core at allocation of corpuscles of a dust from gas, general efficiency of a dust separation essentially drops when concentration of slurry answers such concentration at which it loses properties of a Newtonian fluid. As appears from presented in Figures 22.3–22.6 dependences, together with growth of concentration of slurry above a boundary value,

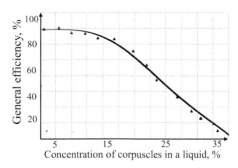

FIGURE 22.6 Dependence of general efficiency on concentration in a liquid of corpuscles of white black.

general efficiency of a dust separation decreases, and the basic contribution to this phenomenon small corpuscles with a size less bring in than 5 microns.

To comment on the dependences presented in drawings, than 5 microns operated with criterion of decrease in efficiency of a dust separation of corpuscles sizes less at the further increase in concentration of slurry at 10% above the boundary. Taking it in attention, it is possible to notice that in case of allocation of a dust of talc growth of concentration of slurry from 36% to 45% calls reduction of general efficiency of a dust separation from 98% to 90% at simultaneous decrease in fractional efficiency of allocation of corpuscles, smaller 5 microns from $\eta = 93\%$ to $\eta = 65\%$.

Analogously for white black: growth of concentration from 7% to 20% calls falling of fractional efficiency from $\eta = 65\%$ to $\eta = 20\%$, for a chalk: growth of concentration from 18% to 30% calls its decrease from $\eta = 80\%$ to $\eta = 50\%$.

Most considerably decrease in fractional efficiency of a dust separation can be noted for difficultly moistened dust – white black (about 50%).

Thus, on the basis of the analysis set forth above it is possible to assert that decrease in general efficiency of a dust separation at excess of boundary concentration of slurry is connected about all by decreasing ability of system to detain small corpuscles. Especially it touches badly moistened corpuscles. It coincides with a hypothesis about updating of an interphase surface. Updating of an interphase surface can be connected also with difficulties of motion of the settled corpuscles of a dust deep into liquids, that is, with viscosity of medium.

The analysis of general efficiency of a dust separation, and, especially, a talcum powder and white black powder specifies that till the moment of achievement of boundary concentration efficiency is kept on a fixed level. In these boundary lines, simultaneously with growth of concentration of slurry dynamic viscosity of a liquid, only this growth grows is insignificant – for talc, for example, to 2.7×10^{-3} Pascal second. At such small increase in viscosity of slurry the estimation of its effect on efficiency of a dust separation is impossible. Thus, it is possible to confirm, what not growth of viscosity of slurry (from 1 to 2.7×10^{-3} Pascal second), and change of its rheological properties influences decrease in efficiency of a dust separation.

The method of definition of boundary extent of circulation of a liquid in impact-sluggish apparatuses is based on laboratory definition of concentration of slurry above which it loses properties of a Newtonian fluid. This concentration will answer concentration of operating fluid, which cannot be exceeded if it is required to secure with a constant of efficiency of a dust separation.

In the conditions of spent research, that is, constant concentration of an aerosol on an entry in the apparatus, and at the assumption that water losses in the apparatus because of moistening of passing air and, accordingly, ablation in the form of drops, is compensated by volume of the trapped dust, the water discharge parameter is defined directly from the recommended time of duration of a cycle and differs for various types of a dust. Counted its maximum magnitude is in the interval 0.02–0.05 l/m³, that is, is close to the magnitudes quoted in the literature.

For periodical regime dedusters this concentration defines directly a cycle of their work. In case of dedusters of continuous act with liquid circulation, the maximum extent of recirculation securing maintenance of a fixed level of efficiency of a dust separation, it is possible to count as the relation:

$$r = \frac{Q_{cir}}{Q_{ir}} \tag{3}$$

where Q_{cir} – the charge of a recycling liquid, m³/h; Q_{ir} – the charge of an inducted liquid on an irrigation, m³/h.

Assuming that all dust is almost completely trapped on a liquid surface, it is possible to write down a balance equation of weight of a dust as:

$$G \cdot (c_{on} - c_{in}) = L \cdot c_{on} \qquad (4)$$

where $(c_{on} - c_{in}) = c_r$ – limiting concentration of a dust, g/m^3.

Then in terms of for calculation of extent of recirculation it is possible to present Eq. (4) formula as:

$$r = \frac{G \cdot c_{on}}{L \cdot c_r} \qquad (5)$$

22.5 CONCLUSIONS

1. Excess of boundary concentration of slurry at which it loses properties of a Newtonian fluid, calls decrease in efficiency of a dust separation.
2. At known boundary concentration c_r it is possible to define boundary extent of recirculation of the irrigating liquid, securing stable efficiency of a dust separation.
3. Magnitude of boundary concentration depends on physical and chemical properties of system a liquid – a solid and changes over a wide range, from null to several tens percent. This magnitude can be defined now only laboratory methods.
4. Decrease in an overall performance of the apparatus at excess of boundary concentration is connected, first of all, with reduction of fractional efficiency of trapping of small corpuscles with sizes less than 5 microns.
5. On the basis of observations of work of an investigated deduster it is possible to assert that change of viscosity of an irrigated liquid influences conditions of a generating of an interphase surface and, especially, on intensity of formation of a drop layer.
6. Selection constructional and the operating conditions, securing high efficiency of a dust separation at small factor of a water consumption, allows to recommend such dedusters for implementation in the industry.

KEYWORDS

- boundary concentration
- cleaning
- fractional efficiency
- gas emissions
- interfacial area
- irrigating liquid
- irrigation
- recirculation
- rheology
- rotoklon
- viscosity

REFERENCES

1. Valdberg, A. J. Wet-type collectors impact-sluggish, centrifugal and injector acts (in Rus.) Moscow: Petrochemistry Publishing House, 1981, 250 pp.
2. GOST 21235-75 (Standard of USSR) Talc and a Talcum Powder the Ground. Specifications (in Rus.)
3. Egorov, N. N. Gas cooling in scrubbers (in Rus.). Moscow, Chemistry Publishing House, 1954, 245 pp.
4. Kutateladze, S. S., Styrikovich, M. A. Hydrodynamics of Gas-Liquid Systems (in Rus.), Moscow, Energy Publishing House, 1976, 340 pp.
5. Kabsch-Korbutowicz, M., Majewska-Nowak, K., Removal of Atrazine from Water by Coagulation and Adsorption. Environ Protect Engineering Journal, 2003, 29(3), 15–24 pp.
6. Kitano, T., Slinna, T. "An empirical equation of the relative viscosity of polymer melts filled with various inorganic fillers." Rheological Acta Journal, 1981, 20(2), 7–14.
7. Margopin, E. V., Prihodko, V. P. Perfection of production engineering of wet clearing of gas emissions at aluminum factories (in Rus.), Moscow, Color. Metal Information Publishing House, 1977, 27–34 pp.
9. Jarzkbski, L., Giowiak. Moscow, Science (Nauka) Publishing House, 1977, 350 pp.
10. Nukijama, S., Tanasawa, Y. Trans. Soc. Mech. Eng. Japan. 1939, V.5.

11. The patent 2317845 Russian Federations, The Rotoklon with adjustable sinusoidal guide vanes. R. R. Usmanova, Zhernakov V.S., Panov A.K. 27 February, 2008.
12. Ramm, V. M. Absorption of gasses (in Rus.). Moscow, Chemistry Publishing House. 1976, 274 pp.
13. Rist, R. Aerosols: Introduction in the Theory (in Rus.), Moscow, World Publishing House, 1987, 357 pp.
14. Stepans, G. J., Zitser, I. M. The Inertia Air Cleaners (in Rus.), Moscow, Engineering Industry Publishing House, 1986, 274 pp.
15. Shwidkiy, V. S. Purification of gasses. Handbook (in Rus.) Ed. by V. S. Shwidkiy, Thermal Power Publishing House, Moscow, 2002, 375 pp.

CHAPTER 23

RHIZOSPHERIC TROPHIC CHAIN: THE ROLE AND STABILITY IN SOIL PROCESSES AND ECOSYSTEMS

N. V. PATYKA,[1] N. A. BUBLYK,[2] T. I. PATYKA,[2] and O. I. KITAEV[2]

[1]NSC "Institute of Agriculture of NAAS of Ukraine," Kiev, Ukraine

[2]Institute of Horticulture, NAAS of Ukraine, Kiev, Ukraine

CONTENTS

ABSTRACT

The authors have considered the problems of rhizosphere interactions in the prism of changing trophic relationships, their resilience in ecosystems. It is shown that the dynamics of rhizosphere interactions with changes in environmental conditions becomes the focus of intensive research and development continuum enrichment area at the roots and the formation of systems of their interaction in the soil profile.

23.1 INTRODUCTION

The composition of the rhizosphere includes the root system of plants and surrounding soil, which is experiencing its effects. This definition is broader than the traditional definition, including the roots and the soil, which is adjacent to them, stressing that this space rhizosphere soil and roots, using the metabolites of roots [4, 11, 20]. The traditional definition ignores important biological interactions with soil biota, as it is, it forms the structure of the soil, which is often regarded as the physical processes, while forgetting about the biologically active complex and the environment.

Rhizosphere is a trophic food webs producers and consumers with a variety of complex organisms (bacteria, micromycetes protozoa and plant root system). There is a definition, which includes features that characterize the formation of biological systems of soil organisms:

1. the use of various energy resources with different speeds for their use;
2. the presence of different life cycles; and
3. variety of habitats.

Trophic interactions in the rhizosphere are necessary to study how complex these submatery working as a unit and have quasi-independent trends. In modern science, there is a clear justification for this kind of research [1, 14, 21], especially when it comes to sets of identical and intertwine dynamics caused by species characteristics, trophic food chains, the level of conservation of the population and rates growth.

Any biological processes can be justified and presented in the form of models that reflect of the complex relationship. We are all familiar with the chemical formulation that gives the primary idea of the basis of the equation of life:

$$6CO_2 + 6H_2O = C_6H_{12}O_6 + 6O_2 \tag{1}$$

Equation (1) includes several concepts. It illustrates the conservation of matter, as each side of the equation is composed of various molecules, but an equal number of atoms and mass. This chemical equation is set as underlying model reflecting rhizospheric interaction. Besides the presentation of photosynthesis and respiration, Eq. (1) represents the relationship between the processes of synthesis and degradation of working at different scales, for example, autotrophic and heterotrophic processes of interaction between plants and herbivores, immobilization of inorganic matter into organic matter and vice versa.

Of course, the biology is not only of carbon, water, hydrogen and rhizosphere basic functions of biological processes lie in the transformation of nutrients such as nitrogen and phosphorus. If nitrogen is added to the equation, while taking into account that the different kinds of microorganisms contained in the biomass of a different ratio of N, P, C, and other elements (different stoichiometry), a similar system of equations and related processes forms the interrelationship of elements in the formation basic necessities of life (Reiners, 1986). When carbon and nitrogen are localized in the organic matter and mineralized into inorganic matter, there are various aspects of formation of bonds from more narrowly focused overground and underground processes to interorganismal by forming biogeochemical pathways immobilization of inorganic metabolites and vice versa. Their same trophic circuitry used to describe trophic interactions [12].

Thus, the use of mathematical modeling (models underlying the rhizosphere functions) necessary for the development of ideas about the stoichiometry and trophic relationships (energy flows) in the rhizosphere. On the one hand, effective models are internally consistent, structurally simple and succinct ideological. On the other hand, the lack of parts, making them interesting biologically, which thus leads to biologically nonlinear inexplicable results. A good example of the latter is unstable mathematical concepts of the theory of mutualistic relationships [15], in contrast to the classical concept of the formation of symbiotic mutualism, which is formed in the rhizosphere.

To get information about mathematically functional orientation in the rhizosphere, it is important to determine the flow of resources in the rhizosphere, which will form a part thereof, and what will be emphasized. What volume of carbon-containing products of photosynthesis exudates produced by plants produced root system during ontogeny plants, which provide the basic framework for the formation of rhizosphere soil. Formation of the rhizosphere caused a rapid and vigorous growth of the root system, which includes twisting of root hairs, death of individual roots and exudation of carbon compounds. The size and dynamics of the formation of the rhizosphere differentiated according to the aboveground biomass, species and ecosystem types. When there is a significant difference in the ratios of contrasting sizes in plants causes may be to limit the ratio of nutrients in the C: N and C: P and selective factor for food resources. A constant ratio and differentiation depending on resource streams greatly simplifies and makes reliable formation of such models.

Carbon fluxes in the rhizosphere help maintain the function of providing the rhizosphere resource necessary for the formation of a mathematical model. Studies rhizosphere indicate that the growth zone of the root system can be divided into zones of a continuum of activity from the root tip to the side chain, where the various microbial groups have access to the downstream root exudates organic [19].

The tip of the root is the lowest root zone. The zone is characterized by the growth of the root rapidly dividing cells and the activity to produce exudates, which also reduces the resistance of the soil. Exudates produced by cells of the roots, provide catering carbon rhizosphere bacteria and fungi, which, in turn, immobilized compounds of nitrogen and phosphorus. At a distance from the zone of root growth nutrient exchange sites are formed, which are formed in the root hairs and a reduction in the level of exudation. Education and death of root hairs defines additional activity and growth of microorganisms. The upper zone of the root system of plants are characterized as areas of remineralization of nutrients that perform predators, as there are regions of mutualistic symbiotic interactions. In each of the areas of the rhizosphere of plants occur production of root system of carbonaceous substances that determine the growth and activity of the rhizosphere microflora [2], protozoa and invertebrates that feed on them [8, 16].

Mathematical description of the trophic food chains in the rhizosphere is based on three kinds of food links [17] – uptake, release of energy and interaction. Each description is based on the view signaling required for their formation.

Description of the relationship based on the study of the rhizosphere of species and changes in their metabolism. Given the extent and size of populations and the interaction of soil microorganisms and the simplest, most of the data presented in the literature, based on the study of trophic inter-actions in soil samples collected volumes of information, research areas, etc. Description provided by relations, despite the volumes of the information received is incomplete and strongly negates the study of diversity and complexity of forming the system.

Chain constant interaction reflects the dynamics influence of one group to another. For example, these descriptions were taken as the basis of several research groups attempted to link the structure of the soil food webs with the transformation of organic matter and mineralization of primary cells [3, 6, 7].

Considering the laws of formation of structure of trophic food chains rhizosphere should be noted that the communication and distribution of energy flows exhibit two patterns, the presence and level of nutrients and biomass within the system that is important for its stability. The first has to do with the flow of energy from the roots and soil biota to higher predators. The constant supply of nutrients in the rhizosphere is a complex process and is determined by several species of dominant microorganisms that function directly in the root system (the so-called complex of plant-microbial energy flows, which occur in the process of differentiation in terms of resource use various physiological and functional scenarios) [7, 9].

Basic food items insects and nematodes, pathogens and microorganisms that symbiotic relationship with the roots of plants is the basis of rhizosphere energy canals. Bacterial energy canal consists of saprophytic bacteria, protozoa, nematodes and some arthropods. Fungi canal energy is largely composed of saprophytic fungi, nematodes and arthropods. Soil bacteria constitute a significant part of the microbial biomass in the rhizosphere, are more efficient in the use of labile root exudates than saprophytic fungi. Unlike bacteria, fungi are more adapted for the use of plant residues. In addition, fungi's and their consumers in the soil occupied by air pores, voids, and live longer. Nutrients within each trophic canal are transformed at different rates, given the characteristics of the substances, which use bacteria and fungi. Coleman [4] found that the differentiation of flows of nutrients due to the formation rate of the food chains because the bacterial energy channel generates a "fast cycle", while the fungi – "slow loop." It is important to note that the mathematical foundation structure of the rhizosphere microbial complex, which forms the system and different dynamic properties, more stable diversity (representation of the group) and complexity (number of links between groups) than in the formation of random (spontaneous) designs.

Modern research has established that the structure of the trophic pyramid is more stable than the alternative structure, such as an inverted pyramid with a higher biomass, supported by higher trophic levels [10, 13].

In addition to the foregoing distribution patterns different biomass and energy flows, coupling constants are asymmetric nature of the interaction, that is, observed beneficial effect for consumers, and vice versa, which depends on the trophic levels. On lower trophic levels, consumers have a strong negative impact on producers and on higher trophic levels producers positively affect consumers. This structuring in close cooperation related to the stability of soil trophic food chains.

Thus, the structure of the interaction forces is closely related to the distribution of biomass and the rate of application of exudates and evaluation are integral components of the interaction forces. Redistribution of the level of interaction simultaneously redistributes biomass and substrate utilization.

Emphasizing the important role of soil biota, functioning and is defined in the rhizosphere of the growth and dynamics of communities, the study of soil as a reservoir of plant nutrients and restrictions are still widespread and are particularly relevant in the consideration and study of the trophic chains, dynamics and development of plant communities. Rhizosphere better studied as a set of individual taxonomic group, and the work not only in the complex, but also has a self-regulating properties (especially when it comes to pool metagenome with similar and interrelated behavior, especially species). Food flows rhizosphere is complex functioning complexes related organisms (bacterial, fungal, root systems and their users). A key distinguishing feature of the complexes is their ability to transform the different types of energy resources at different speeds, characterized by different life cycles and occupy habitats. Organisms within the canal of bacterial and fungal energy responsive to a varying degree of interaction as a unit.

Studies of meadow lands, forest and arctic tundra agricultural systems suggest that the relationship between the energy flows are generally weak on trophic levels associated with bacterial and fungal microflora and

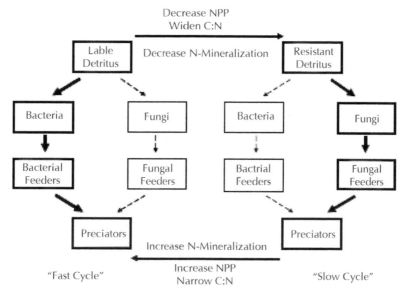

FIGURE 23.1 Generalized bacterial and fungal energy flows in the food web of the rhizosphere [12].

stronger with the simplest. Bonding strength between the energy flow and dominance of a particular depends on the type of ecosystem may vary in certain disorders and controls the flow of nutrients (Figure 23.1).

Using mathematical models, scientists have shown a link between nutrient flow, the level of interaction between communities [5, 12].

Research and models suggest that polymorphism structure is not only important for transfer and storage of nutrients in the system, but also for its homeostasis in general. Modern research has demonstrated the formation of food webs in the rhizosphere as a complex multistage system, which functions as a single organism, which allows to control the flow of nutrients between the producer and the consumer. It is thought that changes in the structure of the basic foundations of the rhizosphere and the rate of passage of nutrient flows are crucial in key processes that shape the ecosystem as a whole. Lack script processes allelopathy destroy the link between the trophic structure, dynamics and transformation proceeds of nutrients, and the lack of homeostasis observed in the rhizosphere under uncontrolled anthropogenic load.

KEYWORDS

- ecosystem
- food web
- rhizosphere
- trophic interaction

REFERENCES

1. Bender, E. A., Case, T. J., Gilpin, M. E. Perturbation experiments in community ecology: Theory and Practice. Ecology, 1984, 65, 1–13.
2. Bringhurst, R. M., Cardon, Z. G., Gage, D. J. Galactosides in the rhizosphere: Utilization by *Sinorhizobium meliloti* and development of a biosensor. Proceedings of the National Academy of Sciences of the USA, 2001, 98, 4540–4545.
3. Brussaard, L., Behan-Pelletier, V. M., Bignell, D. E., (ed.). Biodiversity and ecosystem functioning in soil. Ambio, 1997, 26, 563–570.
4. Coleman, D. C. Energetics of detritivory and microbivory in soil in theory and practice. In: Polis, G. A., Winemiller, K. O. (eds). Food Webs: Integration of Patterns and Dynamics. – Chapman Hall, New York, 1996, 39–50.
5. de Ruiter, P. C., Neutel, A., Moore, J. C. Energetics, patterns of interaction strengths and stability in real ecosystems. Science, 1995, 269, 1257–1260.
6. de Ruiter, P. C., Van Veen, J. A., Moore, J. C., (ed.). Calculation of nitrogen mineralization in soil food webs. Plant Soil, 1993, 157, 263–273.
7. Hunt, H. W., Coleman, D. C., Ingham, E. R., (ed.). The detrital food web in a shortgrass prairie. Biology and Fertility of Soils, 1987, 3, 57–68.
8. Lussenhop, J., Fogel, R. Soil invertebrates are concentrated on roots. In: Keiser, D. L., Cregan, P. B. (eds). The Rhizosphere and plant growth. Kluwer Academic Publishers, Dordrecht, 1991, 111.
9. Moore, J. C., Hunt, H. W. Resource compartmentation and the stability of real ecosystems. Nature, 1988, 333, 261–263.
10. Moore, J. C., de Ruiter, P. C. Invertebrates in detrital food webs along gradients of productivity. In: Coleman, D. C., Hendrix, P. F. (eds). Invertebrates as Webmaster in Ecosystems, CABI, New York, 2000, 161–184.
11. Moore, J. C., McCann, K., Setdldv, H., de Ruiter, P. C. Top-down is bottom-up: Does predation in the rhizosphere regulate aboveground production. Ecology, 2003, 84, 846–857.
12. Moore, J. C., McCann, K., de Ruiter, P. C. Modeling trophic pathways, nutrient cycling, and dynamic stability in soils. Pedobiology, 2005, 49, 499–510.

13. Neutel, A. M., Heesterbeek, J. A. P., de Ruiter, P. C. Stability in real food webs: Weak links in long loops. Science, 2002, 296, 1120–1123.
14. O'Neill, R. V., DeAngelis, D. L., Waide, J. B., Allen, T. F. H. A Hierarchical Concept of the Ecosystem. Princeton University Press, Princeton, 1986.
15. Pimm, S. L. Food Webs. Chapman Hall, London, 1982.
16. Parmelee, R. W., Ehrenfeld, J. G., Tate, R. L. Effects of pine roots on microorganisms, fauna, and nitrogen availability in two soil horizons of a coniferous spodosol. Biology and Fertility of Soils, 1993, 15, 113–119.
17. Paine, R. T. Food webs: Linkages, interaction strength and community infrastructure. Journal of Animal Ecology, 1980, 49, 667–685.
18. Reiners, W. A. Complementary models for ecosystems. American Naturalist, 1986, 127, 59–73.
19. Trofymow, J. A., Coleman, D. C. The role of bacterivorous and fungivorous nematodes in cellulose and chitin decomposition in the context of a root (rhizosphere) soil conceptual model. In: Freckman, D. W. (ed.). Nematodes in Soil Ecosystems. University of Texas Press, Austin, 1982, 117–137.
20. Van der Putten, W. H. L., Vet, E. M., Harvey, J. A., Wdckers, F. L. Linking above- and belowground multitrophic interactions of plants, herbivores, pathogens, and their antagonists. TREE, 2001, 16, 547–554.
21. Yodzis, P. Food webs and perturbation experiments: Theory and practice. In: Polis, G. A., Winemiller, K. O. (eds). Food Webs: Integration of patterns and dynamics. Chapman Hall, New York, 1996, 192–200.

CHAPTER 24

MODELING AND ANALYSIS OF STRESS–STRAIN STATE IN MULTILAYER COATINGS OF ELECTRIC CABLES AT EXTERNAL LOADS

DZHANKULAEVA MADINA AMERHANOVNA

Kabardino-Balkarian State University of H.M. Berbekov, Moscow, Russia

CONTENTS

ABSTRACT

This article presents the results of the calculation of the stress-strain state by the finite element method for single-layer and multilayer electrical cables under the action of external loads. The results of numerical experiments obtained with the software package Solid Works, allow to choose the best parameters of coating of the cable networks to improve their durability.

24.1   INTRODUCTION

The finite element method (FEM) has become practically the universal numerical method for solving all problems, allowing mathematical modeling [1].

It is known [2], [3] that the problem of determining the stress-strain state (SSS) in the three-dimensional case reduces to solving a system of differential equations in partial derivatives of the form

$$G\Delta u_i + (\lambda + G)\frac{\partial \theta}{\partial x_i} + X_i = 0, \quad i = 1, 2, 3,$$

$$\Delta = \frac{\partial^2}{\partial x_1^2} + \frac{\partial^2}{\partial x_2^2} + \frac{\partial^2}{\partial x_3^2}, \lambda = K - \frac{2}{3}G > 0, \tag{1}$$

$$\theta = \frac{\partial u_1}{\partial x_1} + \frac{\partial u_2}{\partial x_2} + \frac{\partial u_3}{\partial x_3}.$$

where $u_i = u_i(x_1, x_2, x_3)$ – displacements of points of a continuous medium under the action of loads, K, G – constants that characterize the elastic properties of medium, $\theta = \varepsilon_{11} + \varepsilon_{22} + \varepsilon_{33}$, X_i – the mass forces.

For the system (1) is constructed functional [3]

$$\Phi(u, \upsilon, w) = \iiint\limits_V \frac{K\theta^2 + 2Ge_{ij}e_{ij}}{2}dV - \iiint\limits_V (Xu + Y\upsilon + Zw)dV \tag{2}$$

$$- \iint\limits_{S_2} (\sigma_1^0 u + \sigma_2^0 \upsilon + \sigma_3^0 w)dS,$$

for which Eq. (1) are the Euler-Lagrange equations. In the Eq. (2) e_{ij}– deviator of the strain tensor, σ_i^0 – given stress on the surface S_2.

Minimum of the functional (2) is defined by FEM with using tetrahedral finite elements.

FEM is the basis of the more parts of modern software systems designed for calculations of computer constructions. In one of these complexes, namely in Solid Works, were created virtual models of polymer coating of the cable networks for solving a class of practically important problems. Consider one of them. Required to determine the stress-strain state of multilayer electrical cables under the action of external loads, such as uniform pressure.

Created in Solid Works virtual constructs were: one –model of a homogeneous (single-layer) cable sheath, the other – a three-layer. Both wares have the identical dimensions: length was 10 cm, the outer diameter was 1 cm, and internal – 0,5 cm. With respect to these objects was modeled situation of compression (in this example the force of 100 N was applied to the outer boundary).

Virtual models of spatial structures have been divided into tetrahedral finite elements (Figure 24.1).

Uniform coating was divided into 6706 finite element (11381 nodes), a three-layer – by 8032 (12559 nodes). When splitting three-layer coating used thick enough finite element mesh, which led to an increase in the number of nodes to calculate SSS.

In the solution the posed problem by finite element method following results were obtained.

Figure 24.2a shows that the stress in a uniform coating is gradually increased closer to the inside boundary. For three-layered cable (Figure 24.2b) maximum stress is observed in the inner layer but, unlike the single-layer shell, no increase is not observed on the whole interval, and reaches its smallest value to the center of the second layer. Despite the essential difference of stress distribution graphics the figures show that their maximum and minimum values are practically identical for both designs.

The Figure 24.3 shows the graphs the distribution circular deformations depending on the radius shells electrical cables.

Of Figure 24.3a shows that the deformation is nearly constant at the border, to which force was applied, and, starting from the center of a homogeneous

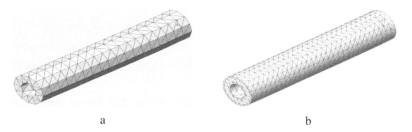

a   b

FIGURE 24.1 Tetrahedral finite element mesh for (a) a single-layer and (b) a three-layer coatings.

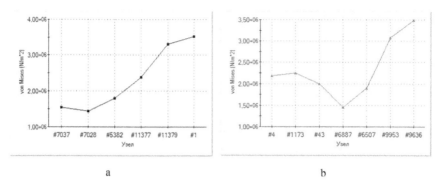

FIGURE 24.2   Circular stress distribution along the radius for (a) a single-layer and (b) a three-layer coatings.

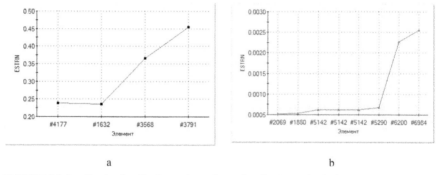

FIGURE 24.3   Strain distributions along the radius for (a) a single-layer and (b) a three-layer coatings.

coating increases linearly. On Figure 24.3b, which shows the change in annular strain in a three-layer shell, seen a sharp increase when approaching the inner boundary of the coating. From this we can conclude that the greatest deformation takes place in the inner layer, while the first two layers will be deformed slightly.

Figure 24.4 shows the graphs moving depending on the radius under the considered loads. Seen from the figures that the three-layer coatings numerical values of moving are significantly less than in homogeneous shells.

Thus the results of numerical analysis show that when the action of external loads, such as, hydrodynamic pressure three-layer polymeric coating of the electrical cable more workable and durable with compare the homogeneous shell.

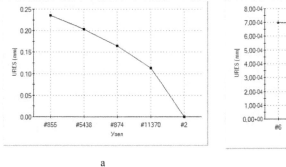
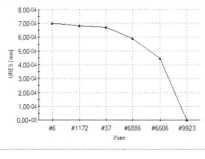

a   b

FIGURE 24.4 Displacement distributions along the radius for (a) a single-layer and (b) a three-layer coatings.

KEYWORDS

- **finite element method (FEM)**
- **multilayer electrical cables**
- **single-layer electrical cables**
- **stress–strain state**

REFERENCES

1. Zenkevich, O. The finite element method in the technique. Moscow: "Mir" ("Peace" in Rus.) Publishing House, 1975, p. 539.
2. Timoshenko, S. P., Goodier, J. Theory of Elasticity. Moscow: "Nauka" ("Science", in Rus.) Publishing House, 1975, p. 574.
3. Oshhunov, M. M., Nagoev, Z. V. Mathematical models of deformable media for intelligent systems virtual prototyping. Nalchik: Publishing house KBSC RAS, 2013, p. 201.
4. Gul, V. E. The structure and strength of the polymers. Moscow: "Khimiya" ("Chemistry", in Rus.) Publishing House, 1978, p. 328.

CHAPTER 25

DEFINITION OF BLOOD SERUM ANTIOXIDANT ACTIVITY OF PATIENTS WITH LIVER PATHOLOGY BY TWO CHEMILUMINESCENT METHODS

N. N. SAZHINA,[1] I. N. POPOV,[2] and G. LEVIN[2]

[1]*Emanuel Institute of Biochemical Physics Russian Academy of Sciences, 4 Kosygin Street, 119334 Moscow, Russia, E-mail: Natnik48 s@yandex.ru*

[2]*Research Institute for Antioxidant Therapy, 137c Invalidenstr., 10115 Berlin, Germany, E-mail: ip@antioxidant-research.com*

CONTENTS

ABSTRACT

An assessment of the total antioxidant activity (AOA) of blood serum of patients with the liver disease by two chemiluminescent methods having different models of free radical oxidation: "Hb-H_2O_2-luminol" and "ABAP-luminol" is carried out. The comparative analysis showed not high correlation of results (r = 0.798), that is explained, mainly, by a distinction of free radical initiation mechanisms and influence of some blood serum components (proteins and bilirubins) on initiation process. More strongly it is shown in model with "Hb-H_2O_2." In this regard, more preferable in clinical practice for an AOA assessment it is necessary to consider model "ABAP-luminol." A comparison of antioxidant parameters of blood serum of patients with the affected liver with some general clinical blood characteristics, such as the content of uric acid, total and direct bilirubin, albumin, parameters of lipid metabolism is carried out.

25.1   AIM AND SCOPE

The comparative analysis of the total blood serum antioxidant activity (AOA) of patients with liver pathology at parallel measurements by two chemiluminescence devices with various models of free radical oxidation for the aim of more suitable model choice in clinical practice. Comparison of antioxidant and some general clinical parameters of blood serum for patients with liver pathology are carried out to establish a correlation between these characteristics.

25.2   INTRODUCTION

Definition of the total AOA of person blood serum is an important task for medico-biological research as AOA determines the protection of person organism in fight against an oxidative stress. Blood is represented difficult substance for research, antioxidant composition which is caused, first of all, by availability in it of amino acids, uric acid, vitamins E, C, hormones, enzymes, and also the intermediate and final metabolism products. [1]. The total AOA is the integrated value characterizing possibility of combine antioxidant action of all blood plasma components taking into account their potential synergism. During 25–30 last years attempts of creation of techniques for total AOA definition at the same time all inhibitors of free radical reactions, which are present in blood plasma are made. However, the role of each inhibitor can significantly differ at various ways of activation of oxidizing processes. Therefore, the choice of adequate systems of AOA assessment has paramount value for the correct interpretation of the received results within clinical laboratory diagnostics [1–3]. Definition of AOA assumes not only detection of one or several substances, and identification of their "functional" activity that can be made in suitable oxidizing system. The main components of any test system for blood serum AOA definition are: the system of radical generation and molecule target which being exposed to oxidation, changes the registered physical and chemical properties. Informational content of the received results depends on a choice of these objects. Now chemiluminescent (CL) methods of blood serum AOA definition are widely used [1, 2]. They are rather sensitive, operative and allow controlling oxidation kinetics directly. One of main distinctions of CL methods is the way of free radicals generation. It can be carried out on chemical or physical-chemical principles (e.g., at interaction of gem-containing derivatives with hydrogen peroxide, at thermodestruction of azo-compounds and radiation of photosensitizes) [2–4].

One of main organs in antioxidative system of an organism is the liver, its synthetic and secretor activity: it belongs as to endogenous antioxidants, so to synthesis of components of inflammatory and anti-inflammatory syndromes [2, 5]. In the liver occur many vital metabolic processes, resulting in the formation and enter in blood substances necessary for the organism, including various endogenous antioxidants, which primarily include uric acid [6] and bilirubin and biliverdin [7]. The liver is the "repository" and some exogenous antioxidants, such as ascorbic acid, which is a synergist for many bioantioxidants and exerts its activity in important for the body's "moments" of oxidative stress [8]. Therefore, studying of blood serum antioxidant properties for patients with liver pathology is an important aspect for understanding of the occurring phenomena in organism protective system.

In the present work of the comparative analysis of the total blood serum antioxidant activity patients with liver pathology at parallel measurements by two chemiluminescence devices with various models of the free radical oxidation is carried out. The contribution to the total antioxidant activity of blood serum of its individual biochemical analytes was assessed also.

25.3 EXPERIMENTAL PART

Blood serum samples of 16 patients with liver pathology (atrophic cirrhosis, liver new growths, etc.) and 18 donors with necessary clinical blood indicators were transferred for research by Myasnikov Institute of clinical cardiology. Measurements of the total AOA in parallel by two CL devices were carried out in Emanuel Institute of biochemical physics RAS.

In the first model of free radical oxidation the system "hemoglobin (Hb) – hydrogen peroxide (H_2O_2) – luminol" in which generation of radicals by Hb and H_2O_2 interaction was used, and luminol plays a role of a chemiluminogenic oxidative substance. Distinctive feature of this model from other oxidation models is that the formed in it radicals can initiate free radical oxidation reactions in vivo as blood contains Hb and H_2O_2. The detailed measurement technique of this model for studying of blood serum AOA and its separate components is given in [4]. The kinetics and detailed scheme of reactions proceeding at "Hb and H_2O_2" interaction are rather difficult. The estimated scheme of the reactions leading to generation of luminol oxidation radical initiators is given in Figure 25.1.

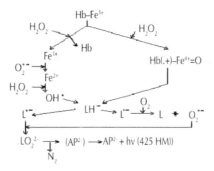

FIGURE 25.1 The estimated scheme of reactions in the system "Hb-H_2O_2-luminol" [4] (LH⁻ – luminol anion, L·⁻ – luminol radical, O2·⁻ - superoxide anion, LO_2^{2-} – luminol-endoperoxide, AP²⁻ and (AP²⁻)* – aminophthalate anion in the basic and excited states, respectively).

As appears from this scheme, H_2O_2 and metHb (Hb-Fe^{+3}) – the oxidized Hb form, making the main part of the commercial Hb preparations, can induce luminol oxidation on two main mechanisms [9]. So, on the one hand, interaction of H_2O_2 with metHb is accompanied by gem destruction and exit from it of iron ions, which participate in education of OH$^{·\,-}$ radicals. Besides, as a result of this interaction active ferril-radicals (Hb($^{·+}$)–Fe^{4+}=O) are formed. Being formed radicals initiate the luminol oxidation which sequence of reactions is well-known now [10]. In the oxidation process L$^{·-}$, $O_2^{·-}$-radicals are formed, a luminol-endoperoxide LO_2^{2-}, and further an aminophthalate anion in excited state (AP^{2-})* upon which transition to the main state light quantum hv is highlighted. Introduction of antioxidants in "metHb-H_2O_2-luminol" system leads to change of CH kinetics and increase in the latent period. Advantage of model: all reagents are available and aren't toxic. Restrictions: instability of H_2O_2 and need of frequent control of its concentration. For realization of this method in the present work the device "Lum-5373" (OOO"DISoft", Russia, www.chemilum.ru) was used according to detail technique [4].

In the second model of the thermo-initiated CL (TIC) luminol as an oxidation substance is used so [2, 3]. Initiation of free radicals happens at thermal decomposition of water-soluble R-N=N-R azo-compound 2.2 '-azo-bis (2-amidinopropane) dihydrochloride (ABAP). In the luminol oxidation processes in the presence of oxygen ROO$^·$- radicals, $O_2^{·-}$- radicals and further LO_2^{2-} and (AP^{2-})* are formed as in the first model. The corresponding scheme of reactions is given below (Figure 25.2) [2]. An advantage of this model – the constant speed of peroxide radical initiation at a stable temperature during a long time. In the water environment at pH = 7.4 and temperature 37°C, radical generation speed Ri, (mol/l)/s = 1.36×10^{-6} [ABAP], where [ABAP] – concentration of an ABAP [11]. A restriction – use of toxic chemical azo-compound. The TIC recordings were carried out with the device "minilum" (ABCD GmbH, Germany) (www.minilum.de) at a temperature of 37 ± 0.01°C.

In both CL systems the key measured parameter for determining of the total water-soluble blood serum component AOA (ACW – "integral antiradical capacity of water soluble compounds") is the latent period. It decides as time from the oxidation initiation moment to a point of

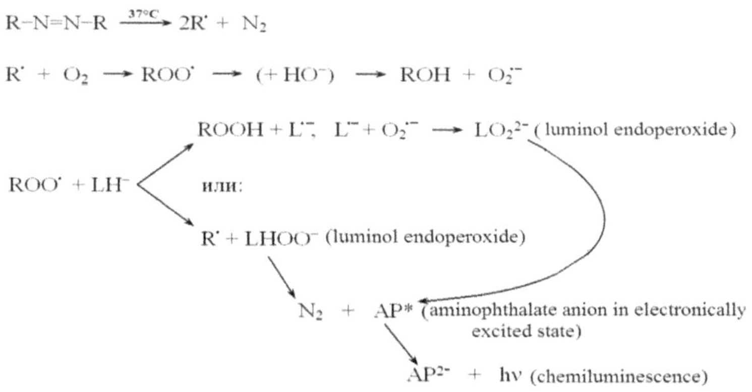

$$R-N=N-R \xrightarrow{37^{\circ}C} 2R^{\cdot} + N_2$$

$$R^{\cdot} + O_2 \longrightarrow ROO^{\cdot} \longrightarrow (+HO^-) \longrightarrow ROH + O_2^{\cdot-}$$

$$ROOH + L^{\cdot-},\ L^- + O_2^{\cdot-} \longrightarrow LO_2^{2-}\ (\text{luminol endoperoxide})$$

$$ROO^{\cdot} + LH^- \Big\langle \quad \text{или:}$$

$$R^{\cdot} + LHOO^-\ (\text{luminol endoperoxide})$$

$$N_2 + AP^*\ (\text{aminophthalate anion in electronically excited state})$$

$$AP^{2-} + h\nu\ (\text{chemiluminescence})$$

FIGURE 25.2   The estimated scheme of reactions in the model "ABAP – luminol" [2].

intersection on an axis of time of the tangent attached to CL-curve in the point corresponding to a maximum of its first derivative dI/dt (Figure 25.7). Calibration of devices was carried out on ascorbic acid, and the total AOA of water-soluble components (ACW) was expressed in the equivalent ascorbic acid content in one liter of blood serum (μmol/L). The measurement error of this parameter for the first device made no more than 20%, for the second didn't exceed 5%.

25.4   RESULTS AND DISCUSSION

In Figure 25.3, kinetic CL-curves received when using both oxidation models for tests of patients are given: test №1-with the lowest value of total AOA (ACW), №2 – with an average, №3 – with the highest. In the first model (Figure 25.3a) for different tests it is characteristic not only change of the latent period, but also considerable change of luminescence intensity while in the second model (Figure 25.3b) only the latent period significantly changes. Especially it is characteristic for patients with the raised content of bilirubin (test №3). For the first oxidation model influence of protein components of blood serum is especially strongly expressed. As shown in Ref. [4], they generally and suppress luminescence intensity.

In Figure 25.4 comparison results of the total AOA (ACW) of water-soluble components of blood serum of studied patients received by both methods are presented. Results show a wide spacing of ACW values: from

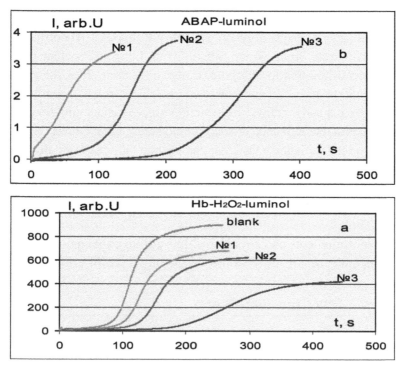

FIGURE 25.3 CL kinetics for "Hb – H$_2$O$_2$ – luminol" model (a) (blood serum volume v=2μL (№1), v=1,5 μL (№ 2,3)) and for the "ABAP-luminol" model (b) (v = 2 μL (№1,2,3)). №1 – test with the lowest ACW value from the donor, test №2 and №3 – from recipients with an average and with the highest ACW. On ordinate axis – intensity of CL (I).

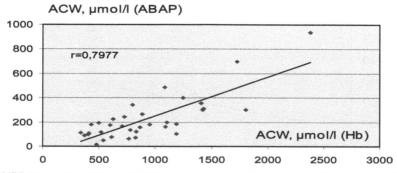

FIGURE 25.4 Correlation of ACW measurement results for donors and recipients (ACW – "integral antiradical capacity of water soluble compounds") for model with hemoglobin (ACW, μmol/L (Hb)) and with ABAP (ACW, μmol/L (ABAP)); r –correlate coefficient.

300 to 2400 μmol/L for the first model, and from 15 to 940 μmol/L for the second (in this case at norm about 100–500 μmol/L). Abnormally high ACW values were observed for patients with the raised content in serum of the general bilirubin to 600 μmol/L above (at norm 1.7–20.5 μmol/L). Low correlation of ACW values for used models (correlation coefficient r = 0.7977) is explained, mainly, by distinction of free radical initiation mechanisms and possible influence of some blood serum components (especially proteins) on the initiation speed that can change the observed latent period considerably. Especially it concerns the first model in which one of radical initiators is very active OH·-radical reacting with many components of blood serum, in particular with proteins [2], which "distract" it from reaction with luminol.

Similar arguments can be adduced and for an explanation of the different dependences of ACW values on the blood serum uric acid content

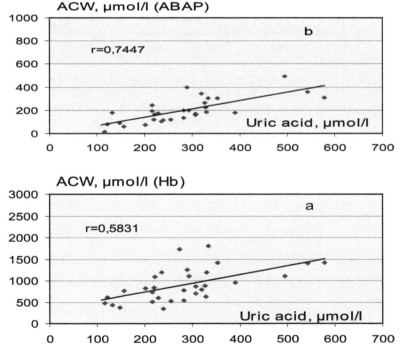

FIGURE 25.5 Dependence of blood serum ACW for donors and recipients on the uric acid content (the biochemical analysis) for oxidation model "Hb-H_2O_2-luminol" (a) and "ABAP-luminol" (b).

(the biochemical analysis) received for these models (Figure 25.5). The relative "stoichiometric" coefficient for uric acid in "ABAP-luminol" system makes about 2.0 [12], while in "Hb-H_2O_2-luminol" – less 1.0 [4]. Interacting with various radical intermediates, uric acid influences in the first model not only on the latent period, but substantially and on the CL intensity. Correlation of ACW with the uric acid content is much worse for the first model (r = 0.583, Figure 25.5a), than for the second (r=0,745, Figure 25.5b). The values of similar correlation coefficients defined with uric acid on "ABAP-luminol" model in other works [12, 13] are rather close to coefficient received in ours experiments. In work [2] using "ABAP-luminol" model on the basis of blood serum measurements for 45 donors it is shown that the contribution of uric acid to ACW makes 64%, and proteins of 5%.

It was shown [14] that in systems of the photo-sensitized (PCL) and thermo-initiated (TIC) CL native amino acids and albumin don't possess anti-radical activity, but get it in processes of oxidizing modification. In "Hb-H_2O_2-luminol" system their share makes about 50% of ACW [4]. Therefore, measurement results are subject to influence of the serum albumin content which at patients with liver pathology is sharply underestimated owing to violation of its synthesizing ability (Figure 25.6). Average value of albumin made 25.21 ±.5.33 g/L (min = 14.4 g/L, max = 37.1 g/L) at norm of 34–48 g/L. The difference of albumin values for different patients reaches 100% and above. It is an essential hindrance at determination of the ACW parameter and calls into question its informational

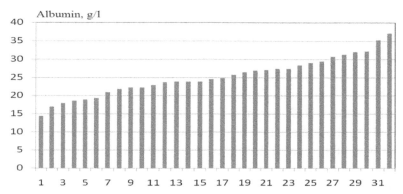

FIGURE 25.6 The albumin content in 32 studied blood serum tests (the biochemical analysis).

content in respect of diagnostics or control of treatment efficiency. In our opinion, the only possibility to use "Hb-H_2O_2-luminol" method in the clinical purposes is the serum deproteinization. However, thus a possibility of quantitative assessment of oxidative stress degree is lost. It is shown in system of TIC by comparison of anti-oxidizing protection with extent of oxidizing damage of blood serum proteins [2].

Thus, a comparative analysis of the total antioxidant activity of blood serum water-soluble components for patients with liver disease (ACW), performed on two free radical oxidation models, showed a relatively low correlation of results (r = 0,798). This is due mainly to the difference in the mechanisms of free radical initiation and the possible impact of some blood serum components (especially proteins) on the process and the rate of initiation. Stronger this effect is manifested in the model " Hb-H_2O_2", where an active OH·-radical-initiator reacts with a number of serum components. The discrepancy in measurement results significant for patients with abnormally high content of certain blood serum components which are differentially inhibit the luminol oxidation due to side reactions. In this regard, more preferred for clinical use to estimate the AOA should be considered the oxidation model with ABAP initiator. Therefore, for further study the correlations of antioxidant and some general clinical parameters of blood serum for patients with liver pathology was chosen the device "minilum" with this model.

Using the appropriate reagents (kits) (ABCD GmbH, Germany www.minilum.de) to determine the activity of fat- and water-soluble antioxidants, antioxidant parameters caused antiradical properties of uric (UA) and ascorbic (ASC) acids and high protein (ARAP – anti-radical ability of proteins) were measured.

Figure 25.7 shows kinetic TIC curves from the series of analyzes. The right curve corresponds to ACW for patient with hyperbilirubinemia, the left – the result of sample re-measurement after its preincubation with the urate oxidase enzyme. The difference of the latent periods corresponds to the contribution of antiradical capacity of uric acid (UA) in ACW. It should be noted that the normal contribution of UA in ACW are 40–80%. In this case its only 9.6%. It indicates on the predominant contribution of bilirubin in ACW for this patient, which was confirmed by the results of the laboratory analysis: total bilirubin was 608.5 µmol/L (normal

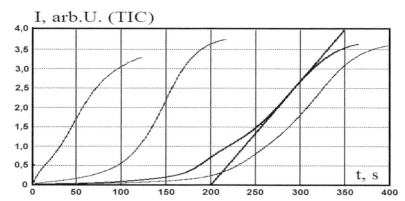

FIGURE 25.7 Kinetic TIC curves, obtained in the measurement of serum ACW for three patients and contribution UA (uric acid) in ACW (explanation in the text).

1.7–20.5), direct – 387.2 μmol/L (normal 0–5.1). The middle curve in Figure 25.7 corresponds to the ACW, which lies within the normal range. Left curve was obtained by analyzing samples from a single donor. The extremely low value of the latent period, which coincides with the blank sample, indicates the complete absence of antioxidant protection. In this case it is necessary to control ACW of donors and take measures to protect organ intended for transplantation. This is necessary to preserve its viability and preventing damage during storage. Efficient use of ascorbic acid for this purpose has been demonstrated previously [15] and repeatedly confirmed in recent years [16, 17].

During ACW measurements it was found that ascorbic acid is absent in blood serum of almost all recipients, since pre-incubation of samples with ascorbate oxidase did not change the character of TIC curves and latent period values. This fact can be explained by "homogeneous" group of seriously ill patients with similar pathology. Similar results were also observed in some cases for less severe diseases [2].

The data in Figure 25.8 show that for most patients the ACW values practically correspond to UA values. The small difference (2.5%) is explained by the contribution of ARAP parameter in ACW due to the action of thiol protein SH-groups. This indicates that the total antioxidant capacity of endogenous hydrophilic antioxidants (ACW) for these patients is mainly determined by the capacity of uric acid. For two patients with abnormally high levels of total and conjugated bilirubin ACW significantly exceeded

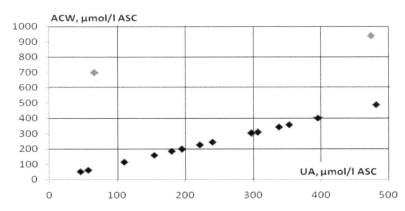

FIGURE 25.8 Comparison of the total blood serum antioxidant capacity (ACW) with antioxidant capacity (UA) of uric acid. Two deviated points – samples of recipients with pronounced hyperbilirubinemia.

the UA. Which forms of bilirubin: direct (bilirubin-glucuronide), indirect (unconjugated pigment in the bloodstream transporting albumin) or free is a major contributor to the ACW, not yet determined. There is evidence that all three forms can be antioxidant activity, but preference is given to indirect bilirubin. It is believed that it prevents oxidation of other ligands for albumin, especially fatty acids, in the complete absence of the reduced form of ascorbic or uric acids [18].

Earlier (Figure 25.5b) the comparative results for serum ACW of patients with their uric acid content were presented. Significant variations in the ACW values are observed. This can be explained by the fact that the significance, which may not correspond to its content. Uric and ascorbic acids may inactivate sequentially two free radicals parameter UA does not reflect the uric acid content in the blood serum, but its antioxidant capacity, its functional remaining in the fully oxidized or semi-oxidized form.

One of the specific tests of liver pathology as parenchymatous organ is the content of the blood serum bilirubin – a bile pigment. Bilirubin is formed by hydrolysis in the spleen splenocytes of a tetrapirroll – heme of hemoglobin. The hydrolysis products are hydrophobic, and unconjugated bilirubi n delivers the transport protein albumin in the liver. Attempts to find in our study the interconnection between the antioxidant parameters of water and lipid phases of blood serum and different forms of biliru-bin, parameters of lipid metabolism, free and bound cholesterol, free fatty acids, triglycerides and phospholipids revealed no significant results. In respect of bilirubin, given its known role in extra- and intracellular lipid and protein protection from oxidation [19] it should be noted that a significant increase in its level in the blood forms the "protection" of blood serum proteins from damage, resulting in low values of ARAP. However, significant correlation between bilirubin in all its forms and parameters ARAP and ACW in the group of studied patients studied in this work not revealed.

Pathology of the liver in the acute stage is accompanied by the loss of his body many vital functions, such as, for example, the synthesis of serum albumin (it was in the normal range only for two patients, for rest patients it was significantly underestimated). The liver function in maintaining the antioxidant homeostasis is disturbed. This function consists, on the one hand, in the storing of ascorbic acid and its release into the bloodstream as required, and on the other hand – in regulating the synthesis of uric acid as an endogenous antioxidant complementary to ascorbic acid. If for any reason (enhanced load by xenobiotics, viral infection, etc.), the reaction of the liver is not sufficient to suppress the oxidative stress caused by these factors, then "wakes up" evolutionarily ancient mechanism of antioxidant protection, manifested in the activation of the heme oxygenase enzyme and the production of bilirubin. This phenomenon is also observed in the intact liver under strong oxidative stress, and with a lack of ascorbic acid in preterm infants, and for patients with severe inflammation [20, 21]. Thus for an effective antioxidant defense may be sufficient even small (nanomolar) concentrations of bilirubin [22, 23]. Evolutionary reason for displacement of this antioxidant defense type is a toxicity of bilirubin at high concentrations [24]. It is noted however that recipients with high preoperative bilirubin levels in liver transplant are more favorable for postoperative period compared with patients with low content in the preoperative period [25].

25.5 CONCLUSIONS

1. The comparative analysis of the total antioxidant activity of blood serum water-soluble components for patients with liver disease (ACW), performed on two free radical oxidation models: "Hb-H_2O_2-luminol" and

"ABAP-luminol" showed, that more preferred for clinical use should be considered the oxidation model with ABAP initiator.

2. Measurement results of the antiradical capacity of blood serum and its some individual components for recipients with liver disease showed significant disturbance of liver function in maintaining the antioxidatet homeostasis.

3. At the liver pathology, the absence of exogenous ascorbic acid and expressed human antioxidant defense, uric acid and bilirubin compensatory become in vivo the major hydrophilic antioxidants.

4. The increased hyperuricemia at different pathological processes can be considered as a activation test of the biological reaction of an inflammation and syndrome of anti-inflammatory compensatory protection.

5. It should be possible to lower hyperuricemia and compensatory function of uric acid in pathological processes by long-time use of optimal doses of ascorbic acid.

ACKNOWLEDGMENT

We thank professor V.N. Titov from Myasnikov Institute of Clinical Cardiology of Russian cardiologic R&D production complex of Russian Minzdrav, doctor S.A. Solonin and professor M.A. Godkov from Sklifosofskiy research Institute of emergency medical care for providing blood serum samples and biochemical analysis.

KEYWORDS

- antioxidant activity
- blood serum
- chemiluminescence
- free radical oxidation
- liver

REFERENCES

1. Bartosz, G. Total antioxidant capacity. *Adv. Clin. Chem.* 2003, 37, 219–292.
2. Popov, I., Lewin, G. Antioxidative homeostasis, its evaluation by means of chemiluminescent methods. *In: Handbook of chemiluminescent methods in oxidative stress assessment.* Transworld Research Network, Kerala, 2008, 361–391.
3. Popov, I., Lewin, G. Antioxidant system of the body and the method of the thermoinitiated chemiluminescence to quantify its condition. *Biofizika.* 2013, 58 (5), 848–856. (in Russian).
4. Teselkin Yu.O., Babenkova, I. V., Lyubitsky, O. B., Klebanov, G. I., Vladimirov Yu.A. Inhibition of oxidation of luminol in the presence of hemoglobin and hydrogen peroxide by serum antioxidants. *Questions of medical chemistry.* 1997, 43(2), 87–93.
5. Titov, V. N. Endogenous oxidative stress system confrontation. The role of DHEA and oleic fatty acid. *Uspekhi sovremennoy biologii.* 2009, 129 (1), 10–26. (in Russian).
6. Glantzounis, G. K., Tsimoyiannis, E. C., Kappas, A. M., Galaris, D. A. Uric acid and oxidative stress. *Curr Pharm Des.* 2005, 11 (32), 4145–4151.
7. Nakagami T, Toyomura K, Kinoshita T, Morisawa, S. A beneficial role of bile pigments as an ednogenous tissue protector: anticomplement effects of biliverdin and conjugated bilirubin. *Biochem Biophys Acta.* 1993, 1158 (2), 189–93.
8. Popov, I., Lewin, G. Antioxidative homeostasis: characterization by means of chemiluminescent technique. In: Packer, L., ed. *Methods in enzymology.* New York. Academic Press; 1999, 300, 437–456.
9. Puppo A, Halliwell, B. Formation of hydroxyl radicals from hydrogen peroxide in the presence of iron. Is hemoglobin a biological Fenton reagent?. *Biochem, J.* 1988, 249(1), 185–190.
10. Faulkner, K., Fridovich, I. Luminol and lucigenin as detectors for $O_2^{\cdot}$. *Free Radiic. Biol. Med.* 1993, 15 (4), 447–451.
11. Niki, E. Free Radical Initiators as Source of Water- or Lipid-Soluble Peroxyl Radicals. *Methods in enzymology.* Eds. L. Packer & A. N. Glazer. New York. Academic Press; 1990, 186, 100–108.
12. Uotila, J. T., Kirkkola, A. L., Rorarius, M. et al. The total peroxyl radical-trapping ability of plasma and cerebrospinal fluid in normal and preeclamptic parturients. *Free Radic. Biol. Med.* 1994, 16 (5), 581–590.
13. Lissi, E. A., Salim-Hanna, M., Pascual, C., Castillo, M. D. Evaluation of total antioxidant potencial (TRAP) and total antioxidant reactivity from luminol-enchanced chemiluminescence measurement. *Free Radic. Biol. Med.* 1995, 18 (2), 153–158.
14. Popov, I., Lewin, G. Photochemiluminescent detection of antiradical activity. VI. Antioxidant characteristics of human blood plasma, low density lipoprotein, serum albumin and aminoacids during in vitro oxidation. *Luminescence;* 1999, 14, 169–174.
15. Popov, I., Gäbel, W., Lohse, W., Lewin, G., Richter, E., Baehr, R. V. Einfluss von Askorbinsäure in der Konservierungslösung auf das antioxidative Potential des Blutplasmas während der Lebertransplantation bei Minischweinen. *Z. Exp. Chirurgie.* 1989, 22, 22–26.

16. Wang, N. T., Lin, H. I., Yeh, D. Y., Chou, T. Y., Chen, C. F., Leu, F. C., Wang, D. and Hu, R. T. Effects of the Antioxidants Lycium Barbarum and Ascorbic Acid on Reperfusion Liver Injury in Rats. *Transplantation Proceedings.* 2009, 41, 4110–13.

17. Adikwu, E., Deo, O. Hepatoprotective Effect of Vitamin C (Ascorbic Acid). *Pharmacology and Pharmacy.* 2013, 4, 84–92.

18. Hunt, S., Kronenberg, F., Eckfeldt, J., Hopkins, P., Heiss, G. Association of plasma bilirubin with coronary heart disease and segregation of bilirubin as a major gene trait: the NHLBI family heart study. *Atherosclerosis.* 2001, 154, 747–754.

19. MacLean, P., Drake, E. C., Ross, L., Barclay, E. Bilirubin as an antioxidant in micelles and lipid bilayers: Its contribution to the total antioxidant capacity of human blood plasma. *Free Rad. Biol. Med.* 2007, 43, 600–609.

20. Fereshtehnejad, S. M., Bejeh Mir, K. P., Bejeh Mir, A. P., Mohagheghi, P. Evaluation of the Possible Antioxidative Role of Bilirubin Protecting from Free Radical Related Illnesses in Neonates. *Acta Medica Iranica;* 2012, 50(3), 153–163.

21. Patel, J. J., Taneja, A., Niccum, D., Kumar, G., Jacobs, E., Nanchal, R. The Association of Serum Bilirubin Levels on the Outcomes of Severe Sepsis. *J. Intensive Care Med.* 2013, 28(3), 230–236.

22. Dore, S., Takahashi, M., Ferris, C. D. et al. Bilirubin, formed by activation of heme oxygenase-2, protects neurons against oxidative stress injury. *Proc. Natl. Acad. Sci.* 1999, 96(5), 2445–2450.

23. Baranano, D. E., Rao, M., Ferris, C. D., Snyder, S. H. Biliverdin reductase: a major physiologic cytoprotectant; *Proc. Natl. Acad. Sci.* 2002, 99(25), 16093–16098.

24. Ames, B. N., Cathcart, R., Schwiers, E., Hochstein, P. Uric acid provides an antioxidant defense in humans against oxidant- and radical-caused aging and cancer: a hypothesis. *Proc. Natl. Acad. Sci.* 1981, 78, 6858–6862.

25. Igea, J., Nuno, J., Lopez-Hervas, P. et al. Evaluation of delta bilirubin in the follow-up of hepatic transplantation. *Transplant. Proc.* 1999, 31(6), 2469.

INDEX

S

T